Smart Cities and Construction Technologies

Edited by Sara Shirowzhan
and Kefeng Zhang

Published in London, United Kingdom

IntechOpen

Supporting open minds since 2005

Smart Cities and Construction Technologies
http://dx.doi.org/10.5772/intechopen.86103
Edited by Sara Shirowzhan and Kefeng Zhang

Contributors
Faham Tahmasebinia, Anqi Shi, Sara Shirowzhan, Samad M.E. Sepasgozar, Alireza Kaboli, Mohammad Mojtahedi, Laurence Kimmel, Jack Peacock, Reyhaneh Karimi, Behnaz Avazpour, Alan McNamara, Farhad Soheili, Aman Kumar, Jasvir Singh Rattan, Mohsen Ghobadi

First published in London, United Kingdom, 2020 by IntechOpen
IntechOpen is the global imprint of INTECHOPEN LIMITED, registered in England and Wales, registration number: 11086078, 7th floor, 10 Lower Thames Street, London, EC3R 6AF, United Kingdom
Printed in Croatia

British Library Cataloguing-in-Publication Data
A catalogue record for this book is available from the British Library

Additional hard and PDF copies can be obtained from orders@intechopen.com

Smart Cities and Construction Technologies
Edited by Sara Shirowzhan and Kefeng Zhang
p. cm.
Print ISBN 978-1-83880-199-1
Online ISBN 978-1-83880-200-4
eBook (PDF) ISBN 978-1-83880-398-8

An electronic version of this book is freely available, thanks to the support of libraries working with Knowledge Unlatched. KU is a collaborative initiative designed to make high quality books Open Access for the public good. More information about the initiative and links to the Open Access version can be found at www.knowledgeunlatched.org

For EU product safety concerns:
IN TECH d.o.o., Prolaz Marije Krucifikse Kozulić 3, 51000 Rijeka, Croatia, info@intechopen.com or visit our website at intechopen.com.

Meet the editors

Dr Sara Shirowzhan is a lecturer of City Analytics at the Faculty of Built Environment, University of New South Wales, Australia. She completed her PhD in Geomatic Engineering at the School of Civil and Environmental Engineering, University of New South Wales, Australia. Sara started teaching in built environment disciplines from 2007 and currently teaches GIS, BIM, and major project courses at undergraduate and postgraduate levels. She has supervised 22 Master's students and has been involved in university and government funded research projects. She has developed her skills in advanced technologies for sustainable and smart built environments such as laser scanning, nD BIM, nD GIS, GIS based app and dashboard creation, digital twins, big data analysis, VR, and AR.

Dr Kefeng Zhang is a senior research associate at Water Research Centre, School of Civil and Environmental Engineering, UNSW Sydney. Dr Zhang received his PhD from Monash University in 2015 and developed a framework for the validation of micropollutant removal in passive stormwater treatment systems. He has been working mainly in the area of urban stormwater management, including stormwater quality monitoring and modelling, green technologies (e.g., stormwater bio-retention systems and green walls) that are developed under the concept of Water Sensitive Urban Design (also called Sponge City, Low Impact Development, and Natural-based Solution), as well as development of integrated urban water modelling.

Contents

Preface

Smart technologies are increasingly used in different businesses and industries. A wide range of these technologies and digital innovations are applicable to smart cities. In addition, the construction industry also uses the advantages of disruptive technologies to enhance productivity and safety. At some point, cities, infrastructures, and construction processes can share the data collected at different stages, and help to increase sustainability, decrease emissions, and improve efficiency. These concepts refer to a broad range of subtopics, practices, and technologies that cannot be collected by a book or journal. Therefore, this book is limited to two sections, which include nine chapters covering some of the subtopics that will be useful for scholars in the field of smart cities and construction technologies. Each section will be described in the following two paragraphs, and details of each topic are available in the full chapters.

Section 1, Smart City Studies

This section begins with a review chapter on a journey from conventional cities to smart cities. In this chapter, changes of conventional cites to smarter cities through improvements of residents' lifestyles and reduction of waste and pollution are explored. Another chapter discusses earth scrapers as an alternative building method for underground city development that requires less surface area and has lower heating and cooling costs. An analysis of thermal comfort seems promising as it shows internal temperature of such buildings ranging from 20 to 38 degrees, which is within the comfort zone temperature range. In another chapter, the structure of floating cities bridge is analysed to determine the amount of load, stress, and deflection of the structure. The floating bridge is defined as a connector between cities where the population is high and there is significant land scarcity for urban developments. In another study, a GIS based risk and safety analysis is conducted for designing smarter buildings in the future. The study presents a set of innovative evaluation criteria to assess the perceived safety in educational buildings. The feelings of safety are mapped within a GIS environment to identify the most problematic locations in terms of perceived safety. In a different chapter, effective factors in students' sense of attachment to school and educational efficiency are explored. Additionally, omission of certain open spaces such as courtyards in cities has been explored and identified as factors contributing in the reduction of residents satisfaction in their apartments so the factors affecting the satisfaction related to the open spaces in apartments are discussed for enhancement of the quality of life, health, and wellbeing in future smart cities.

Section 2, Technology Applications in Construction

In Section 2, construction technologies such as intelligent construction contracts, 5D BIM for 3D printing, and adoption of Virtual Reality are discussed. Automated intelligent contracts (i.e. "iContracts") as the protocols for facilitation, verification, or enforcement of the negotiations are proposed for minimising the need for conventional human management. In the era of smarter cities, the construction industry is also transforming the processes to digital construction.

Building Information Modelling (BIM) plays a key role in digital construction. Dimensions of BIM are being introduced to make calculations and visualisations more effective and efficient. The fifth dimension of BIM, known as 5D BIM, can be defined as BIM based cost estimation and it is currently very important for quantity surveyors in construction processes. The application of 5D BIM for 3D printing of building and infrastructure components is discussed in this section. Virtual Reality (VR) is another emerging technology being increasingly used in construction and smart cities. While the application of this technology is proven for training, education, and safety modellings, the adoption of VR in construction is still questionable and one of the chapters of this book explores this.

Sara Shirowzhan
Faculty of Built Environment,
University of New South Wales,
Sydney, Australia

Kefeng Zhang
Water Research Centre,
School of Civil and Environmental Engineering,
University of New South Wales,
Sydney, Australia

Section 1

Smart City Studies

Chapter 1

A Journey from Conventional Cities to Smart Cities

Aman Kumar and Jasvir Singh Rattan

Abstract

In the 1990s, all the cities were conventional cities because at that time there was no concept of the Internet. In the last decade, the concept of smart city has been spread all over the world with the advancement of Internet, and it puts an impact on both small and large cities. In the present scenario, the urban areas affected by various problems and smart cities are only the solutions. The definition of smart cities depends on the word of "smart," which means a digital city, intelligent city, and sustainable city. The basic concept of the smart cities is that the city should be "green," more "accessible," and more "liveable." This chapter explores the changes made in the conventional cities to become the smart cities, which helps to improve the lifestyle of city people and reduce waste and pollution.

Keywords: smart cities, smart environment, digital city, sustainable urban development, conventional cities

1. Introduction

On the inception of civilization, when man decided to settle and abandon a wandering lifestyle, his immediate requirement was a suitable shelter. Firstly, he lives

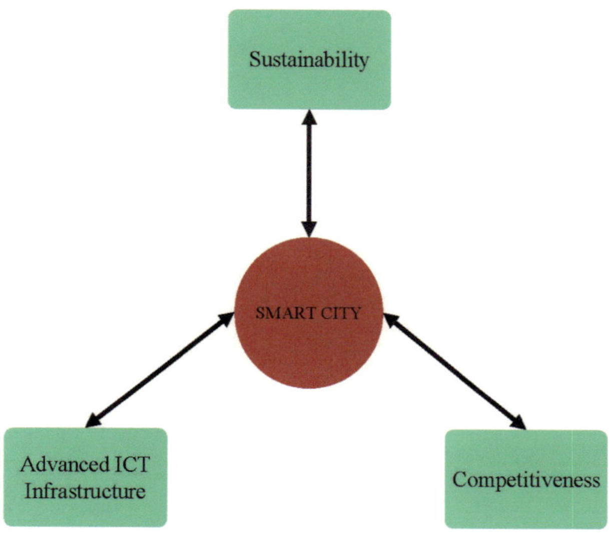

Figure 1.
Main goals of Small City projects.

in caves, then shifted to man-made huts, from man-made to villages and lastly shifted to conventional cities. The conventional cities defined as the cities in which people live without using the latest technology in all aspects. The latest technology related to the smart infrastructure, smart traffic arrangements, and smart living standards of life. But in present scenario, with the fast growth in population, create challenges to the government as well as public sector. So, smart cities provide the best solution in such cases for the better life of urban areas. The first smart city of the world, which

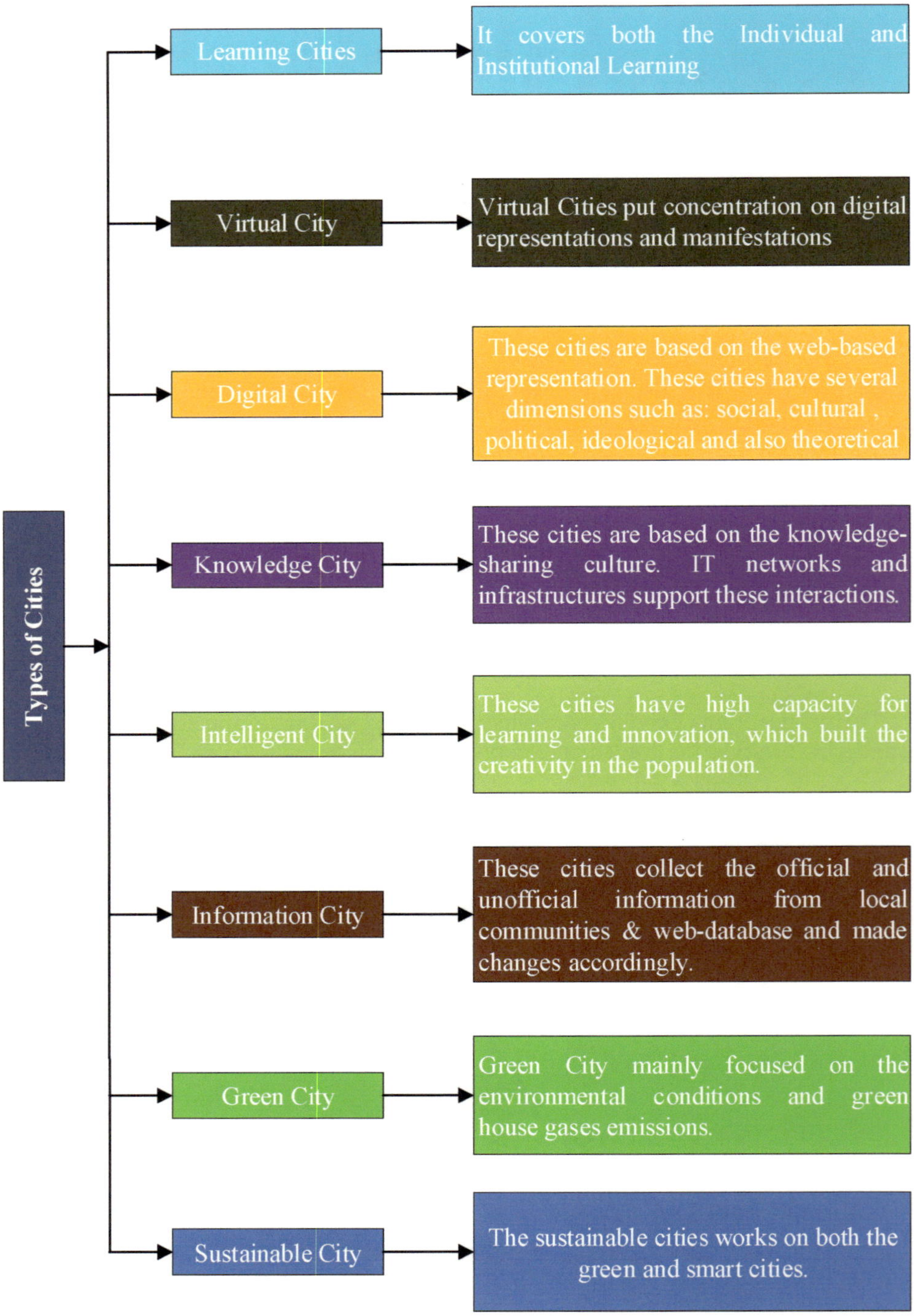

Figure 2.
Different types of cities.

was introduced in the year of 2014 is in Seoul the capital of South Korea [1, 2]. The strategic goals of Seoul smart city projects are Sustainability, Advanced Information and Communication Technology (ICT) and Competitiveness as shown in **Figure 1** [3]. The developing country like India, the question is not that what to build, but where to build. The land resources are very limited and population growth at a very fast rate. In the case of Singapore from the independence of 1965, they reclaim the land about 23% using the sea. As the urban populations rise the world will face unprecedented challenges, like house, transport, and food. In the future as the rural population decrease ensuring food stability will be key. Singapore adopts good way to grow the food in the urban environment [4]. Traditional land farming grows produce outside on a flat linear area. But sicinius system grows food inside in any building expanding agriculture's footprint up into the sky. This is called vertical farming. Vertical farming is good for the areas which have land scarcity [5]. As populations age and incidence of diabetes continue to across the developed world, so, it is important to encourage the citizens to eat smart and well [6]. The main objectives of this chapter are: (a) the difference between smart cities and conventional cities (b) artificial intelligence in smart cities (c) various definitions of smart cities and (d) smart city indexed parameters (**Figure 1**).

2. Literature review

2.1 Types of cities

The various types of cities which include Learning, Cities, Virtual City, Digital City, Knowledge City, Intelligent City, Information City, green City and Sustainable City are defined in **Figure 2**.

2.2 Definitions of smart cities

The various definitions of the smart cities have been described in **Table 1**, which are taken from the various research papers from the origin of the smart cities concept.

Reference	Definition
Hall [7]	In smart cities the author Hall mainly focused on the good condition of building and infrastructures with the addition of security of city people.
California Institute [8]	The California institute describes the smart city on the basis of smart community. The smart community is that community which is ready to work with the latest technology.
Caragliu et al. [9]	Caragliu defined the smart city on the foundation of human being, smart capital and modern IT infrastructure which are the fuel of sustainable city and also provide high quality life to the urban civilization.
Su et al. [10]	Smart City is the product of Digital City combined with the Internet of Things.
Batagan [11]	A Smart City well-performing built on the "smart" combination of endowments and activities of self-decisive, independent and aware citizens.
Lu et al. [12]	Smart City is a city in which it can combine technologies as diverse as water recycling, advanced energy grids, and mobile communications in order to reduce environmental impact and to offer its citizens better lives.
Dameri [13]	AAuthor defined the smart cities on the basis of geographical area in which he include various parameters such as: quality of life, intelligence development, use of green and eco-friendly resources and able to state the rules and policy for the city government and its development.

Table 1.
Definitions of smart cities from literature.

3. Conventional cities

These cities have very intricate systems of organized people, business, "transportation", "communication networks", "services", and "utilities" [14, 15]. As the city grows, they create the technical, social, economic and organizational pressures that put economic and environmentally sustainability in jeopardy. The conventional cities use the non-renewable resources of energy. In these cities, there is no proper management for garbage and control pollution [3]. The typical type of conventional city shown in **Figure 3**.

In conventional cities the people are suffering from many problems such as water issues, insufficient of electricity, poor traffic management, poor lifestyle [16], conventional education system, the hospitals are not well equipped, lack of employment, safety, and security, people are not much aware of the facilities and rule and regulations and growth of uncontrollable population [17]. To overcome these all the problems, the conventional cities take the transformation to the smart cities.

Figure 3.
Image of conventional city which is situated in India.

4. Smart city

The smart cities can be defined as "it is the future of the urban population/cities, made safe and secure and having good environment" [18]. The material used for the construction of smart cities should be eco-friendly and the city having tracking and decision making algorithms [19]. Smart cities demand carefully planning at early ages, it is important the city will fulfill the requirements of government and citizen. A clear strategy must address two key factors: "functions" and "purposes," the function refers to aesthetical appearance and operations of a city, and "purposes" refers to the benefits promised by a smart city model [20].

5. Background of smart cities

With the increase in the population and the growth of urban cities, cities have been endowed unprecedented power on politics, economic and technology and play a crucial role in the world. As reported in history, every global crisis related to finance or energy always triggers a technical revolution, and the winner will lead the economic development as shown in **Figure 4** [21]. Traditional city development pattern based on straightforward use of non-renewable sources such as coal-fuel, petroleum, and natural gas, which results in pollution and creates a problem for

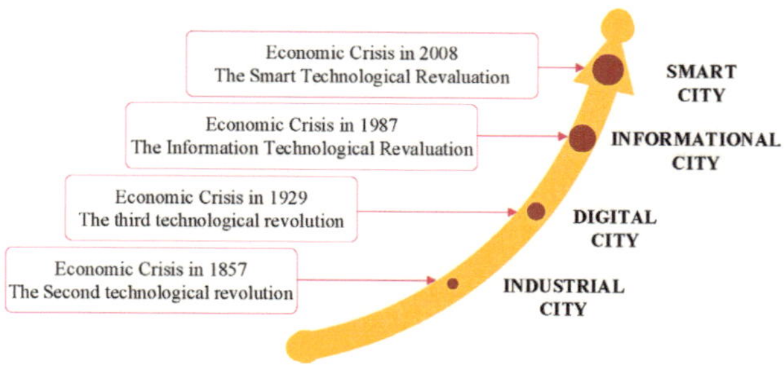

Figure 4.
Background of smart cities.

people who work in that area. The "sustainable development", "Smart Growth", and "Intensive Development" have been proposed by the government mostly in all the countries [22]. The concept of the smart city has been proposed by the IBM company at the end of 2009 following "smart earth" and it receives attentions of governments, scholars, and citizens.

In developing countries, the word "smart" is related to the mental ability of the people to understand the technology and use it in a right way. To prepare the urban population in such a way, so they can deals with the problems smartly and find the perfect solution in an eco-friendly way. The major problem in the urban cities is energy crises. So, in smart cities foremost focus of the government is that to promote the quality of urban operations and the policies of the urban development [23].

5.1 Basic requirement for a city

The basic components for every city are Land, People, Government, and Infrastructure.

Land: Land is the same for both cities conventional and smart cities. The land is that area, where the city is to be growing.

People: The people in conventional cities are different from smart cities. In conventional cities, people are not much aware of the latest technological features. But in smart cities, people are aware of smart technology such as the Internet of Things (IoT), wireless communication and cashless technology.

Government: Support of the government is very important and it is the responsibility of the government to introduce the latest technology in the smart cities and also provide awareness among the citizens.

Infrastructure: Now in this present era, the land space is very limited, and it is essential to place the infrastructure in such a way so every citizen has comfortably sufficient space for their living.

5.2 Applying the smart cities

Currently, the construction of the smart cities is on-trend, more than 50 counties work on the concept of smart cities. There are many problems in the urban areas such as environmental issues, overpopulation and traditional methods are not able to cope up from that problem [24]. From the last decade, all the countries begin to start working on Internet of Things (IoT), cloud computing, networking and other innovative technologies which helps to make the cities intelligent for the purpose of energy conservation and improving their long-term benefits for future [25]. The world's most top 10 smart cities are London, New York, Amsterdam,

Paris, Reykjavik, Tokyo, Singapore, Copenhagen, Berlin and Vienna reported by IESE Cities in Motion in the year 2019. The representation of these cities with their location is shown in **Figure 5**.

The distribution of the cities is shown on the world map in **Figure 5**. As per indexing of IESE 2019, they indexed 165 smart cities in all over the world depending on their dimensions such as "Economy", "Human Capital", "Social Cohesion", "Environment, Governance", "Urban Planning", "International outreach", "Technology", "Mobility" and "Transportation".

5.3 List of smart cities continent wise

The given table shows the countrywise list of the smart cities which are indexed by the IESE smart system 2019. This table categorized the countries according to their continent with their ranks. In Asia the Tokyo city of Japan placed in first rank as per the smart city index and in Australia the Sydney city

Figure 5.
Top 10 smart cities in the world (indexed by IESE smart system).

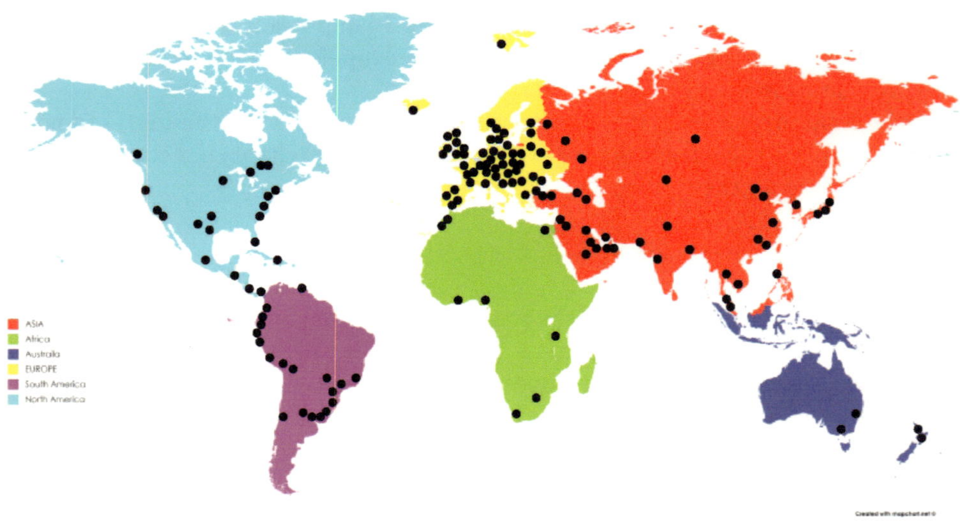

Figure 6.
Geography distribution of smart cities (continental wise).

of Australia placed in the first rank. Similarly, the other cities with their country name placed rankwise, indexed in **Table 2** [26]. And the geographical distribution of smart cities shown in **Figure 6**.

5.4 The framework of smart cities

The framework of smart cities is divided into six parts such as smart mobility, smart environment, smart people, smart living, smart governance, and smart economy as shown in **Figure 7** [27, 28].

Continent	Cities (countries)
Asia	Tokyo (Japan), Singapore (Singapore), Hong Kong (China), Seoul (South Korea), Taipei (Taiwan), Shanghai (China), Osaka (Japan), Tel Aviv (Israel), Nagoya (Japan), Beijing (China), Dubai (United Arab Emirates), Kuala Lumpur (Malaysia), Bangkok (Thailand), Guangzhou (China), Istanbul (Turkey), Shenzhen (China), Ho Chi Minh City (Vietnam), Jerusalem (Israel), Tbilisi (Georgia), Doha (Qatar), Abu Dhabi (United Arab Emirates), Almaty (Kazakhstan), Baku (Azerbaijan), Ankara (Turkey), Jakarta (Indonesia), Kuwait City (Kuwait), Amman (Jordan), Bangalore (India), Tianjin (Chin), Manama (Bahrain), Mumbai (India), Manila (Philippines), Riyadh (Saudi Arabia), New Delhi (India), Kolkata (India), Lahore (Pakistan) and Karachi (Pakistan).
Africa	Casablanca (Morocco), Tunis (Tunisia), Douala (Cameroon), Cape Town (South Africa), Nairobi (Kenya), Cairo (Egypt), Johannesburg (South Africa), Rabat (Morocco) and Lagos (Nigeria).
Australia	Sydney (Australia), Melbourne (Australia), Wellington (New Zealand) and Auckland (New Zealand).
Europe	London (United Kingdom), Amsterdam (Netherlands), Paris (France), Reykjavik (Iceland), Copenhagen (Denmark), Berlin (Germany), Vienna (Austria), Stockholm (Sweden), Oslo (Norway), Zurich (Switzerland), Helsinki (Finland), Madrid (Spain), Munich (Germany), Barcelona (Spain), Basel (Switzerland), Bern (Switzerland), Geneva (Switzerland), Frankfurt (Germany), Hamburg (Germany), Goteborg (Sweden), Dublin (Ireland), Milan (Italy), Rotterdam (Netherlands), Lisbon (Portugal), Edinburgh (United Kingdom), Prague (Czech Republic), Brussels (Belgium), Dusseldorf (Germany), Cologne (Germany), Stuttgart (Germany), Lyon (France), Eindhoven (Netherlands), Valencia (Spain), Birmingham (United Kingdom), Glasgow (United Kingdom), Tallinn (Estonia), Warsaw (Poland), Bratislava (Slovakia), Antwerp (Belgium), Budapest (Hungry), Vilnius (Lithuania), Rome (Italy), Seville (Spain), Manchester (United Kingdom), Leeds (United Kingdom), Malaga (Spain), Riga (Latvia), Nice (France), Moscow (Russia), Linz (Austria), Palma de Mallorca (Spain), Marseille (France), Duisburg (Germany), Porto (Portugal), Ljubljana (Slovenia), Liverpool (United Kingdom), Wroclaw (Poland), Nottingham (United Kingdom), Lille (France), Zaragoza (Spain), Zagreb (Croatia), A Coruna (Spain), Bucharest (Romania), Murcia (Spain), Athens (Greece), Bilbao (Spain), Florence (Italy), Turin (Italy), Minsk (Belarus), Kiev (Ukraine), Sofia (Bulgaria), Naples (Italy), Belgrade (Serbia), Saint Petersburg (Russia), Sarajevo (Bosnia and Herzegovina), Skopje (North Macedonia) and Novosibirsk (Russia).
South America	Santiago (Chile), Buenos Aries (Argentina), Montevideo (Uruguay), San Jose (Costa Rica), Panama (Panama), Bogota (Colombia), Rosario (Argentina), Rio de Janeiro (Brazil), Brasilia (Brazil), Sao Paulo (Brazil), Mexico City (Mexico), Medellin (Colombia), Santo Domingo (Dominican Republic), Cordoba (Argentina), Lima (Peru), Curitiba (Brazil), Asuncion, (Paraguay), Cali (Colombia), La Paz (Bolivia), Belo Horizonte (Brazil), Guayaquil (Ecuador), Guatemala (Guatemala) and Caracas (Venezuela).
North America	New York (USA), Los Angeles (USA), Chicago (USA), Toronto (Canada), San Francisco (USA), Washington (USA), Boston (USA), Montreal (Canada), Ottawa (Canada), Miami (USA), Phoenix (USA), Dallas (USA), Denver (USA), Philadelphia (USA), Vancouver (Canada), Houston (USA), Quebec (Canada), Seattle (USA), San Diego (USA), San Antonio (USA) and Baltimore (USA).

Table 2.
List of smart cities (continent wise by their ascending order).

Figure 7.
The framework of smart cities.

6. Artificial intelligence for smart cities

Artificial intelligence is an interdisciplinary science and technology founded on many disciplines such as Computer Science, Biology, Psychology, Linguistics, Mathematics, and Engineering [29]. The leader of artificial intelligence John McCarthy defined AI as "The science and engineering of making intelligent machines, especially intelligent computer programs" [30]. Artificial Intelligence is that technology, in which we program the machine in such a way, that they think intelligently and works accordingly human intelligence. AI works on the basis of the human brain, how the human brain thinks, decide to work and find the solution of the problems [31]. In smart cities artificial intelligence plays a crucial role because in

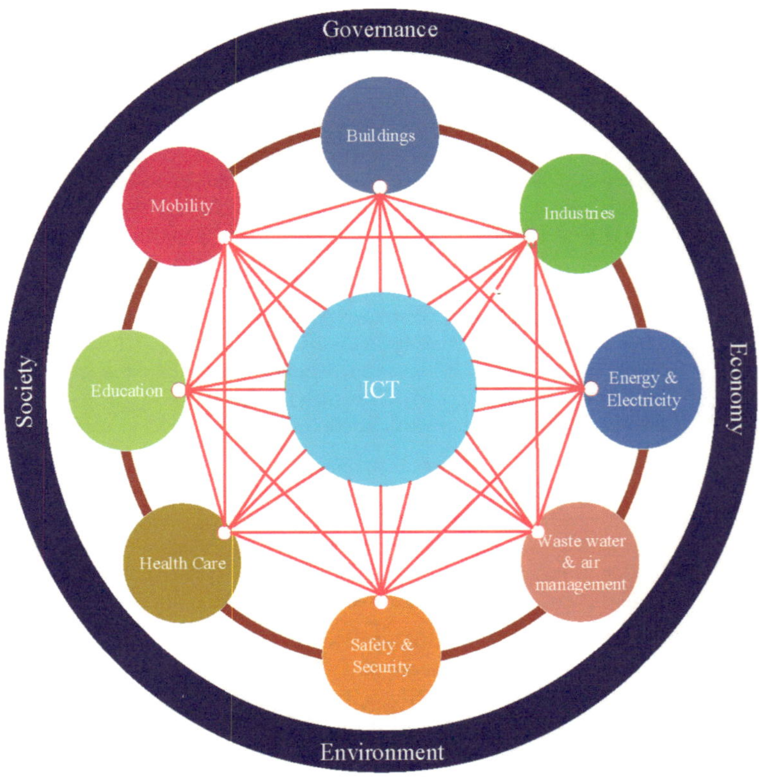

Figure 8.
The infrastructure of smart cities.

smart cities, the data is big so we need big data studies by using artificial intelligence [32]. The infrastructure of smart cities with various parameters shown in **Figure 8**.

7. Smart cities indexed parameters

By IESE Cities in motions, they defined nine parameters, on the bases of these parameters we indexed the smartness of the cities. These nine parameters are: Economy of the city, Human Capital of the city, International relation of the city, Medium of the Transportation inside the city and with connected cities, Environmental conditions, advanced technology, urban planning, governance, interaction of the people and Infrastructure as shown in **Figure 9** [33].

7.1 Economy

The economy plays an important role to make the city smart with their advancement in the public sector. The economic dimension includes; plans of local economic development, initiatives by entrepreneurs and plans by industry.

7.2 Human capital

The main focus of the smart cities to improve the human capital. The human capital includes following parameters such as: higher education, research programs, business school, Museum and art gallery etc, [34]. A smart governance only works if the human capital retaining that talent and promoting creativity.

7.3 International outreach

The international outreach means that the smart city has that brand so that become the favorite place for the tourists. The international outreach also contains the foreign investments and representing the city in global level.

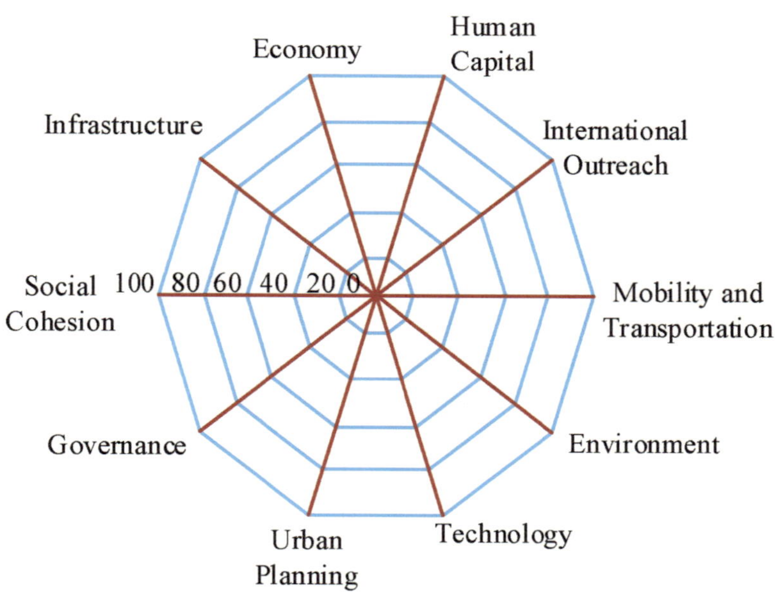

Figure 9.
Smart City indexed parameters.

7.4 Mobility and transport

The future cities are facing two problems in the area of mobility and transport: (a) access to public areas and (b) movement inside the city. The mobility and transport contain both the vehicle and infrastructure parts. So, that in smart cities use those types of vehicle which having low CO_2 emissions. And as per the infrastructure point of view, it should be free and traffic jams and well organized. The main smart cities mobility and transport parameters are: traffic index, metro length, metro stations, flights, high-speed trains and gas stations.

7.5 Environment

The environment should be green, clean and free from all types of pollutions. To clean the environment from the pollution by planting the anti-pollution trees, support green buildings system and use alternative sources of the energy. The indictor of the environment which showing their index in smart cities are: CO_2 emission, renewable water resources, solid waste management and pollution index.

7.6 Technology

The technology is not only the important aspect of the smart cities, but it is the backbone of the cities that make them "smart" [35]. Technology puts an important impact on the smart cities, it increases the quality of life, provides safety and security. The various indicator of technology are social media, smart mobile phone used by city people, WiFi supported city and Camera Security.

7.7 Urban planning

Urban planning mostly contains all types of infrastructure which includes buildings, railway lines, airports or seaports. The indicators of the urban planning are: no. of household uses sanitation facility, high-rise buildings and number of people per household.

7.8 Governance

Governance describes the effectives of the people to solve the city problems. The governance indicators are research centers, embassies, government buildings, government development index and democracy [36, 37].

7.9 Social cohesion

Social cohesion defined as "it is a sociological dimension of the city, as the degree of consensus among the members of a social group or as the perception of the belonging to a common situation or project" [38]. The indicator of the social cohesion is rate of crime, health, hospitals, happiness index, price of property, mortality and terrorism [39].

7.10 Infrastructures

The infrastructure contains all types of building structures, and other city structures. The material used during the construction should be eco-friendly and after destruction, it is useable for other construction purposes (**Figure 9**).

8. Conclusion

In this era, we are mostly dealing with the Internet which is becoming smarter and more intelligent than ever. As per the definitions of the smart city, it should be including efficient water, primary health, smart grids, wireless Internet access, efficient waste, road safety, clean energies, transportation and education, and e-governance. The security of the smart city is very important and it depends mainly on the factor such as city governance, socio-economic and technological dependence. The primary role of these factors to resolve the security-related issues. ICT technology plays an important role to form a smart city. This does not only help to solve the infrastructure-related problems, but also take care of security-related problems. The ultimate mission of the smart city is to brave innovative creativity by the government and commercial sector to initiate the financial progress and develop day -to day quality of life by empowering local development and connecting latest technology to benefits the residents. The key parameters of smart cities are small buildings, smart energy, ease of mobility and personal security. This chapter discusses how conventional cities converted to smart cities using latest technology.

In smart cities, various technologies adopted to make citizen life comfortable and healthy. But it is also the duty of country citizens to maintain that facility properly. The smart city concept is very good and unique, which provide all the basic amenities for urban citizens. The contribution of this chapter in the area of smart cities is that, it provides the various parameters which decide the ranking of the smart cities in all over the world. The major key points are: advancement in the area of artificial intelligence, work with smart technology for smart cities and basic components of sustainable development. The future scopes of the smart cities will be work on "smart citizen" to provide the digital solution in the field of healthcare, education and other government policies.

Conflict of interest

"The authors declare no conflict of interest".

Author details

Aman Kumar[1*] and Jasvir Singh Rattan[2]

1 Indo-Global of Engineering, Mohali, India

2 National Institute of Technical Teachers Training and Research, Chandigarh, India

*Address all correspondence to: aman.civil16@outlook.com

References

[1] Bowerman B, Braverman J, Taylor J, Todosow H, Wimmersperg U. The vision of a Smart City. In: Proceeding of the 2nd International Life Extension Technology Workshop. Paris France; 2000. pp. 1-7

[2] Lehr T. Mint. Smart cities: Vision on-the-ground. In: Smart Cities. Springer;2018. pp. 3-15. DOI: 10.1007/978-3-319-59381-4_1

[3] Camero A, Alba E. Mint: Smart City and information technology. Cities. 2019;**93**:84-94. DOI: 10.1016/j.cities.2019.04.014

[4] Zhou Y. Mint: The path towards smart cities in China: From the case of Shanghai expo. In: Proceeding of the REAL CORPS; 2014. China; 2010. pp. 1023-1027

[5] Hollands R. Will the real smart city please stand up. Cities. 2008;**12**(3):303-320. DOI: 10.1080/13604810802479126

[6] Schuler D. Digital cities and digital citizens. In: Digital Cities II Computational and Sociological Approaches. 2002. pp. 71-85. DOI: 10.1007/3-540-45636-8_6

[7] Hall P. Mint: Creative cities and economic development. Urban Studies. 2000;**37**(4):639-649. DOI: 10.1080/00420980050003946

[8] Cities and Regions in the New Learning Economy [Internet]. 2001. Available from: http://www.oecd.org/internet/citiesandregionsinthenewlearningeconomy.htm

[9] Caragliu A, Bo C, Nijkamp P. Smart cities in Europe. Journal of Urban Technology. 2011;**18**(2):65-82. DOI: 10.1080/10630732.2011.601117

[10] Su K, Lie L, Fu H. Smart City and the applications. In: Proceedings International Conference on Electronics, Communications and Control (ICECC). 2011. pp. 1028-1103. DOI: 10.1109/ICECC.2011.6066743

[11] Batagan L. Mint: Smart cities and sustainability models. Economica. 2011;**15**(3):80-86

[12] Lu P, Chen S, Zheng Y. Mint: Artificial intelligence in civil engineering. Mathematical Problems in Engineering. 2012;**2012**:1-23. DOI: 10.1155/2012/145974

[13] Dameri R. Mint: Searching for Smart City definition: A comprehensive proposal. International Journal of Computers and Technology. 2013;**11**(5):2544-2551. DOI: 10.2429/ijct.v11i15.1142

[14] Ergazakis K, Metaxiotis K, Psarras J. Towards knowledge cities: Conceptual analysis and success stories. Journal of Knowledge Management. 2004;**8**(5):5-15. DOI: 10.1108/13673270410558747

[15] Green Cities Programme [Internet]. 2010. Available from: https://www.oecd.org/regional/greening-cities-regions/46811501.pdf

[16] European Initiative on Smart Cities [Internet]. 2019. Available from: https://setis.ec.europa.eu/set-plan-implementation/technology-roadmaps/european-initiative-smart-cities

[17] Glasmeier A, Chirstopherson S. Mint: Thinking about smart cities. Cambridge Journal of Regions, Economy and Society. 2015;**8**(1):3-12. DOI: 10.1093/cjres/rsu034

[18] Anthopoulos L. Mint: Understanding smart cities: A tool for smart government of an industrial trick. Public Administration and Information Technology. 2017;**22**:2-14. DOI: 10.1007/978-3-319-57015-0

[19] Cosgrave E. Mint: The smart City: Challenges for the civil engineering sector. Smart infrastructure and construction. 2017;**170**(4):90-98. DOI: 10.1680/jsmic.17.00012

[20] Coiclelis H. The construction of the Digital City. Environment and Planning B: Planning and Design. 2004;**31**:5-19. DOI: 10.1068/b1299

[21] Lim C, Kim K, Maglio P. Mint: Smart cities with big data: Reference models, challenges and considerations. Cities. 2018;**82**:86-99. DOI: 10.1016/j.cities.2018.04.011

[22] Trindade E, Hinnig M, Costa M, Marques J, Bastos R, Yigitcanlar T. Mint: Sustainable development of smart cities: A systematic review of the literature. Journal of Open Innovation: Technology, Market, and Complexity. 2017;**3**(11): 1-14. DOI: 10.1186/s40852-017-0063-2

[23] Jain R, Abraham D. Mint: Computational approaches to enable smart and sustainable urban system. Journal of Computing in Civil Engineering. 2019;**33**(6):1-2. DOI: 10.1061/(asce)cp.1943-5487.0000850

[24] Anthopoulos L, Fitsilis P. From digital to ubiquitous cities: Defining a common architecture for urban development. In: Proceedings of Sixth International Conference on Intelligent Environments. 2010. pp. 301-306. DOI: 10.1109/IE.2010.61

[25] Mohmood Z. Smart cities. In: Computer Communications and Networks. Springer; 2018. DOI: 10.1007/978-3-319-76669-0

[26] IESE. Cities in motion [Internet]. 2019. Available from: https://blog.iese.edu/cities-challenges-and-management/2019/05/10/iese-cities-in-motion-index-2019/

[27] Nam T, Pardo T. Conceptualizing Smart City with dimensions of technology, people, and institutions. In: The Proceedings of the 12th Annual International Conference on Digital Government Research. 2011. pp. 282-291. DOI: 10.1145/2037556.2037602

[28] Yigitcanlar T, Kamruzzaman M, Buys L, Ioppo G, Marques J, Costa E, et al. Mint: Understanding 'smart cities'; intertwining development drivers with desired outcomes in a multidimensional framework. Cities. 2018;**81**:145-160. DOI: 10.1016/j.cities.2018.04.003

[29] See S. Mint: Artificial intelligence computing for a Smart City. International Conference on Smart Cities, Infrastructure, Technologies and Applications. 2018;**224**:6-15. DOI: 10.1007/978-3-319-94180-6_2

[30] Wu S, Chen T, Wu Y, Lytras M. Mint: Smart cities in Taiwan: A perspective on big data applications. Sustainability. 2018;**10**(1):1-14. DOI: 10.3390/su10010106

[31] Komninos N. The architecture of intelligent cities, integrated human, collective and artificial intelligence to enhance knowledge and innovation. In: 2nd International Conference ne on Intelligent Environments. 2006. pp. 13-20

[32] Bakici T, Almirall E, Wareham J. Mint: A Smart City initiative: The case of Barcelona. Journal of the Knowledge Economy. 2013;**4-2**:135-148. DOI: 10.1007/s13132-012-0084-9

[33] Galati R. Mint: Funding a Smart City: From concept to actually. In: Smart Cities. Springer; 2018. pp. 17-39. DOI: 10.1007/978-3-319-59381-4_2

[34] Ahmadian E, Sodagar B, Mills G, Byrd M, Bingham C, Zolotas A. Mint: Sustainable cities: The relationship between urban built forms and density indicators. Cities. 2019;**95**:1-15. DOI: 10.1016/j.cities.2019.06.013

[35] Baltac V. Mint: Smart cities- a view of societal aspects. Smart Cities. 2019;**2-4**:538-548. DOI: 10.3390/smartcities2040033

[36] Ruhulandt R. Mint: The governance of smart cities: A systematic literature review. Cities. 2018;**81**:1-23. DOI: 10.1016/j.cities.2018.02.014

[37] Tomor Z, Meijer A, Michels A, Geertman S. Mint: Smart governance for sustainable cities: Findings from a systematic literature review. Journal of Urban Technology. 2019;**26-4**:3-27. DOI: 10.1080/10630732.2019.1651178

[38] Gaffney C, Robertson C. Mint: Smarter than smart: Rio de Janeiro's flawed emergence as a smart. Journal of Urban Technology. 2018;**25-3**:47-64. DOI: 10.1080/10630732.2015.1102423

[39] Miklian J, Hoelscher K. Mint: Smart cities, mobile technologies and social cohesion in India. Indian Journal of Human Development. 2017;**2**(1):1-16. DOI: 10.1177/0973703017712871

Chapter 2

Earthscraper: A Smart Solution for Developing Future Underground Cities

Faham Tahmasebinia, Kevin Yu, Jiachen Bao, George Chammoun, Edwin Chang, Samad Sepasgozar and Fernando Alonso Marroquin

Abstract

This chapter reports on the finite element analysis of the "earthscraper," proposed by BNKR Arquitectura. It was proposed as an alternative building method for the future, as it requires less surface area and lower operating costs than an equivalent aboveground structure. A 2D model of the cross section of the structure was created using Strand7 for steady-state thermal analysis. This solver gave internal temperature ranging between 20 and 38°C between the bottom apex and the surface, respectively. This provides a comfortable temperature by default, displaying the lesser dependency on heating and cooling costs. A 3D model was also created to analyze the effect of lateral earth pressure by the use of the linear static solver. Results give a maximum lateral displacement of 527 mm and 19.8 mm on the exterior and interior walls, respectively. The model was used for earthquake analysis in accordance with AS/NZS1170.4, requiring the natural frequency and spectral response solvers. Twenty-five modal frequencies were found, with 99.6% of the mass of the structure contributing to the direction under analysis. The maximum horizontal displacement of the structure under the designed earthquake loads was found to be 19.2 mm.

Keywords: earthscraper, future cities, underground development, numerical modelling, finite element analysis

1. Introduction

The United Nations has projected that the world population will reach 9.7 billion by 2050, with the increase of 2.5 billion as to date. The increased pressure on housing solutions to accommodate for the growing population while reducing the carbon footprint has led BNKR Arquitectura [1], based in Mexico City, to formulate a radical solution called the "earthscraper."

Such a building is an inverted pyramid with the ability to accommodate 100,000 while using a small surface area, allowing easy implementation into developed cities. In addition, its founding into the soil eliminates any wind loading while also ensuring a constant warm temperature imposed by the surrounding soil, reducing the energy consumption for heating and cooling.

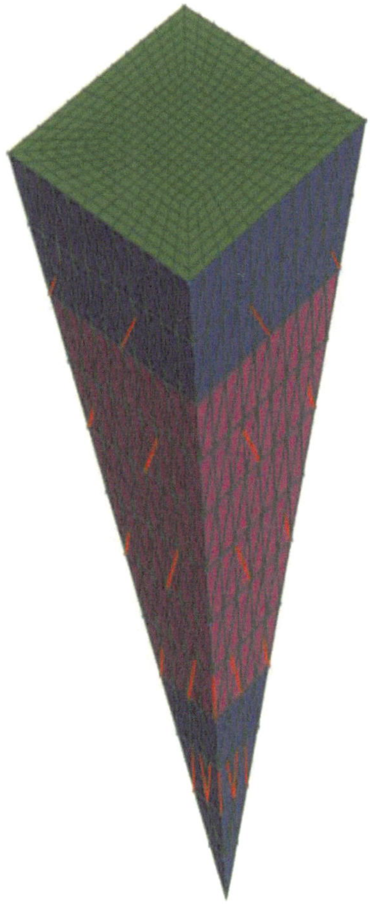

Figure 1.
Conceptual design of earthscraper.

The projected populated rise in Sydney is forecast as between 2 million [2], while urban sprawl is limited by the surrounding Blue Mountains, countryside, and ocean [3] (Clarke, 2016). As such, the main objective of this project will investigate the feasibility of implementing the earthscraper in Sydney's geology and climatology. The theoretical loads imposed by earth pressures, thermal loads, and earthquakes loads will be determined, and the structures response will be modeled using Strand7, a finite element method processor.

Post-processing of the results will give the displacement of the building under each respective load, providing an insight into the expected structural performance. The possibility of such a structure as an addition in the future cities of 2050 shall be determined in this chapter (**Figure 1**).

2. Structural members

Below is the description of the members of the structure:

Floor system: reinforced concrete one-way floor slabs, 0.3 m thick with N16 reinforcement, sitting on steel beams and acting compositely.

Beams: steel beams of varying length extending from bottom of the exterior wall to the top of the interior wall, interconnected to form a truss system. Steel 360UB44.7 beams supporting the concrete slabs.

Details of the structural elements	Suggested structural element size	Suggested design standards
Steel columns	460UB82.1	AS4100 [4]
Reinforced concrete floor system	40 MPa, thickness 0.3 m, N20 rebars	AS3600 [5]
Reinforced concrete interior wall	40 MPa, thickness 2 m, N20 rebars	AS3600 [5]
Reinforced concrete exterior wall	50 MPa, thickness 2 m, N20 rebars	AS3600 [5]
Truss beams	360UB50.7	AS4100 [4]
Perlucor ceiling	Thickness 0.1 m	AS1288 [6]

Table 1.
Structural elements.

Columns: steel UB360.44.7 column spacings at 6.5 m, with floor-to-floor height of 5 m. Steel columns replaced by truss system at locations of intersection.

Walls: reinforced concrete with 1.5–2 m thickness and N20 reinforcement, acting compositely.

Foundation: concrete on rock-concrete pads for columns extruding into the concrete.

Cables: rock bolts to tie back structure.

Ground: water table at 60 m depth, soft soil of unit weight 17 kN/m^3 from depths 0–10 m, and then Hawkesbury sandstone with unit weight 24 kN/m^3 from depths 10–300 m.

Ceiling: perlucor glass ceiling with improved strength and heat insulation compared to normal glass.

Soil: spring dampeners on the exterior of the structure to simulate the soil and the structure's deformations within the soil.

Table 1 shows the general element sizes used in the design of the structure. Element sizes were chosen based on design standards and structural strength, combined with numerical analysis to determine the most appropriate member size for the expected loads.

3. Structural system

Figure 2a shows the dimensions of the structure, with a Perlucor glass ceiling and reinforced concrete walls and a reinforced concrete core running through the center of the structure. The green lines represent the outline of the interior concrete core, the black lines are the outline of the whole structure, the dark blue lines represent the floors, and the light blue lines on the top represent the Perlucor glass.

Figure 3a and **b** shows the interior design of the structure for the first level, with the steel cables forming a truss system between the exterior wall and the interior core to provide buckling and bending resistance from the lateral earth pressure loads and earthquake loads.

Figure 3c shows the spring dampeners with fixed restraints at the end away from the wall and free restraints at the connection with the wall to simulate the behavior of the structure within the soil mass.

The concrete floor slabs use N16 reinforcement bars at a spacing of 100 mm, while the reinforced concrete wall and core use N20 reinforcement bars at a spacing of 100 mm. The reinforcement helps the structure resist tensile loads due to the large bending moments caused by the lateral earth pressure.

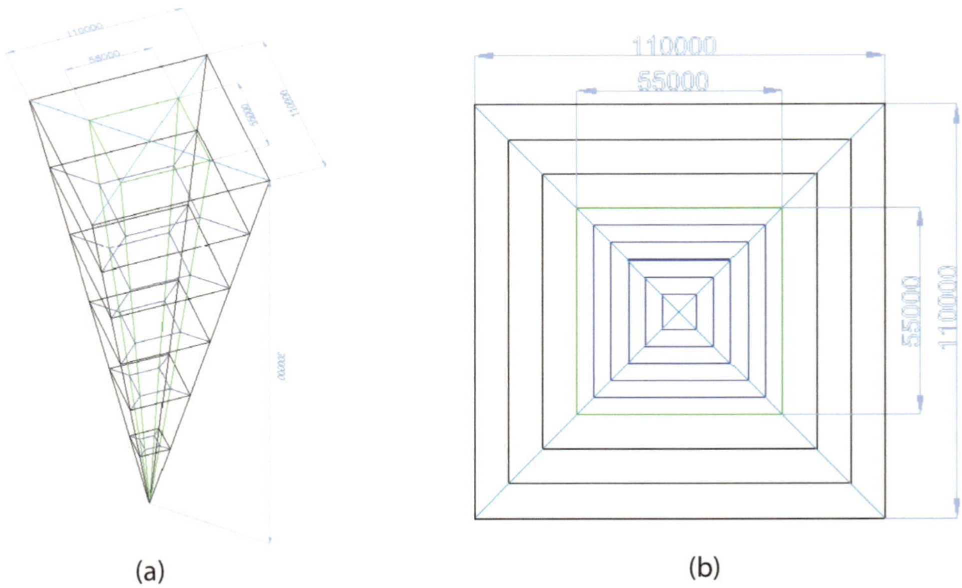

Figure 2.
(a) CAD rotated section view of entire structure and (b) CAD plan view from ground level.

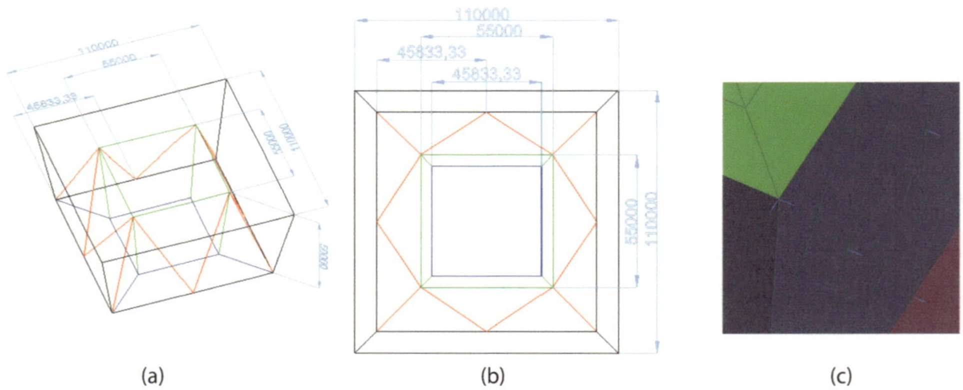

Figure 3.
(a) CAD section view of first level, (b) CAD plan view of first level and (c) spring dampeners.

The system resists vertical loads through load transfer from the roof and floor slabs into the columns, through the beams and truss system. The vertical loads within the columns are then transferred into the surrounding sandstone. The beams and columns of each individual floor have been omitted from the design, and the 60 floors of the 300 m deep structure have been simplified to 6 floors, as the vertical loads are not the focus of this structural design, since the surrounding Hawkesbury sandstone will be able to withstand the dead and live loading of the structure.

The system resists lateral loads with the truss system and floor slabs acting as struts, reducing the overall deflection of the structure and distributing lateral loads throughout the floor and into the surrounding bedrock. The floors resist loads by transferring vertical live and dead loads into the support steel beams, and the slabs resist lateral loads by transferring the loads either into the interior wall or exterior wall. The beams resist loads by transferring horizontal and vertical loads into nearby columns or walls. In addition, the beam truss elements provide different load paths and brace the structure to further reduce the stress concentrations on members. The columns resist load by transferring horizontal and vertical loads into the base pads and then into the surrounding bedrock.

Since the structure and its 50 floors are very complex to design for using Australian design codes, it was simplified into 6 larger floors to consider overall trends of stresses and deflections. This was done to locate areas of significant stresses and deflection and where the structure is most likely to fail. Strategies were then implemented to counteract these problem areas, such as increasing wall thickness or concrete strength.

4. Loads

4.1 Lateral earth loads (earth pressure)

The loads imposed onto the earthscraper by the surrounding soil and water were calculated using effective earth pressures, which are shown below and expanded in Appendix A:

$$\sigma_h'(z) = K_0 \sigma_v'(z), \tag{1}$$

$$K_0 = 1 - \sin(\phi), \tag{2}$$

Eq. (1) shows the relationship between vertical effective pressure and lateral earth pressure through the at-rest coefficient of earth pressure, K_0. Eq. (2) gives an expression for K_0, given by Jaky [7] as a function of the internal friction angle ϕ. These lateral earth pressures act on the exterior wall, causing horizontal deflection of external and internal wall after load transfer.

Figure 4 shows the lateral pressures varying linearly, with a change in gradient at 10 m depth with the change from soft soil to Hawkesbury sandstone. At a depth of 60 m is the water table, introducing pore water pressures and increasing the lateral loading on the structure. This representation of earth pressures.

4.2 Thermal loads

The thermal loads imposed onto the structure include the solar heat flux, the convection currents between the structure and outside fluid, and the external ambient pressure of the surface. The ambient temperature surrounding the earthscraper was determined through research of past studies. The topsoil temperature was averaged across 1980–2006 for the top 20 cm, determined as 19.65°C [8].

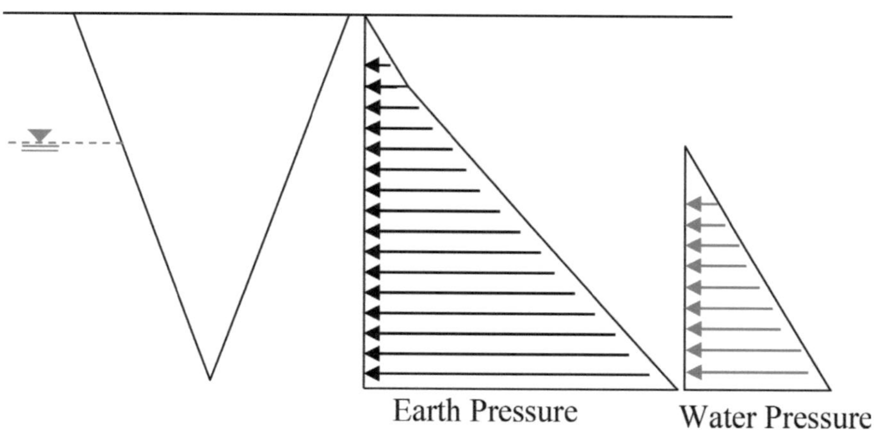

Figure 4.
Vertical profile of lateral earth pressure—a combination of soil pressure and water pressure.

The temperature of the soil is taken to linearly increase with depth, with a determined thermal gradient of 0.07°C/m.

The ambient temperature of the surface was chosen as the mean temperature at the Sydney Observatory Hill Bureau Weather Station since 1859, determined as 22.1°C [8]. However, due to the incoming solar radiation, the heat flux also contributes to the heat transfer to the Perlucor ceiling. Calculations for an absorbing blackbody give a heat flux of 1.37 kW/m^2; however, due to cloud albedo, atmospheric absorption, and radiation reflection, the solar exposure is taken as 69.44 W/m^2 [8].

4.3 Earthquake loads

The response of the structure under earthquake loads requires the use of dynamic analysis, as the earthscraper falls into the earthquake design category III (EDCIII), as determined in Appendix C. This method shall comply with AS/NZS1170.4 [9] Section 7 with the use of a modal response spectrum analysis.

5. Structural design

The initial analysis was undertaken with all concrete wall elements and floors as the same plain concrete with 40 MPa strength and 0.5 m wall thickness. Upon linear static and spectral analysis, it was found that horizontal deflections along the second, third, and fourth floors were much greater than the first, fifth, and sixth floors. As a response, concrete reinforcement was added throughout the structure to better resist the lateral loads, and the concrete wall thickness was increased to 2 m for the second, third, and fourth floors. In addition, the concrete strength used was increased to 50 MPa for the second, third, and fourth floors. The first, fifth, and sixth floors were also increased in wall thickness to 1.5 m.

To further combat the deflections and stresses found within the structure, a steel beam truss system was created to connect the exterior wall to the interior wall, providing additional strutting and load transfer throughout the structure. 360UB44.7 beams were chosen for their relative light weight and strength. Perlucor glass was chosen for the ceiling design for its superior strength and thermal properties compared to standard glass and to fulfill the need to have a transparent ceiling material for light transfer downwards into the structure.

6. Strand7 models

6.1 2D model

A two-dimensional model was created to model the temperature within the structure. A quad 4 plate was created with the thermal properties of air and the dimensions of a vertical cut through the middle of the structure. The structure was then subdivided, and the top layer of plates was changed to the structural properties of glass with the thermal properties of Perlucor. The sidewalls were then converted to the structural and thermal properties of concrete (**Figure 5**).

6.2 3D model

The model was formed by creating tri-3 elements as the exterior and interior walls. The floors and ceiling were then created using quad 4 elements. The truss

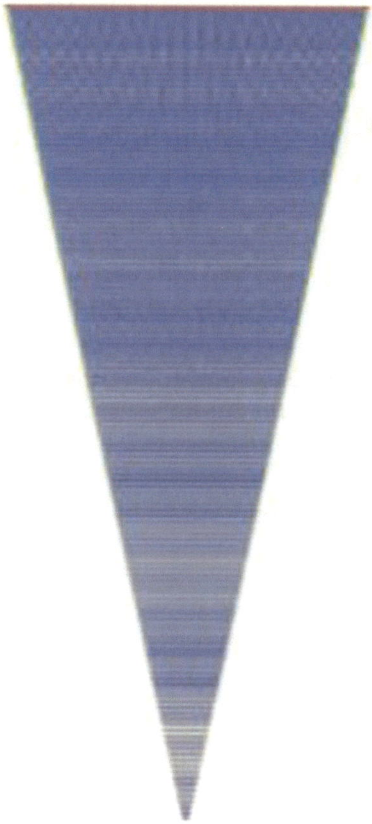

Figure 5.
2D model of the earthscraper created using Strand7 cross section showing the perlucor ceiling (red), green concrete exterior walls (green), and the air inside (blue).

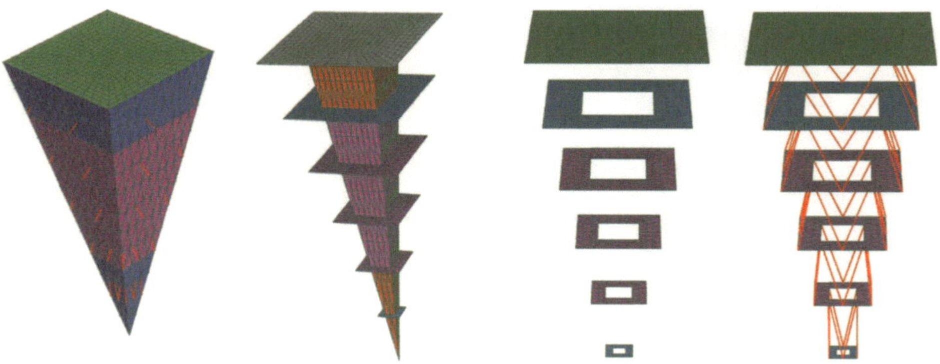

Figure 6.
Stages of 3D model creation—exterior walls, interior walls and floors, and beam truss (from left to right).

beams were then made using beam 2 elements. Each element type was then separated into six groups for their respective floors. The plate elements were then subdivided. Finally, spring dampener beam 2 elements were created by selecting the nodes on each face of the structure and then extruding by increment of 1 m in the negative Z direction or the face's equivalent axis perpendicular to the vertical axis pointing outward. Fixed conditions were then applied to the base node of the structure (tip of the pyramid), as well as the node of the spring dampeners not attached to the structure (**Figure 6**).

7. Numerical analysis

7.1 Thermal analysis

The transfer of thermal energy occurs in two thermal paths. The Perlucor ceiling at the surface ambient temperature absorbs the solar radiation, transporting the heat internally through conduction and heating the internal air through convection. In addition, the walls of the structure also at the ambient temperature of the surrounding soil conduct this energy through the wall system and expel this energy through convection within the structure.

An ambient temperature of 22.1°C was applied at the top surface, and the linearly increasing ambient temperature of the soil was applied along the sidewalls. A convection coefficient of 11 and 23 W/mK was applied along the Perlucor ceiling and sidewall, respectively, and heat flux from the sun of 69.44 W/m^2 was applied along the top surface. The results of the static thermal solver are shown below, alongside the model used.

Figure 7 shows a maximum temperature of 38°C at the bottom vertex, a region designed to harness geothermal energy. This, however, will not be sufficient for energy production according to the thermal analysis. Otherwise, the building reaches thermal equilibrium at comfortable temperatures. Ranging between 19 and 25°C for the upper third of the structure almost eliminates the need for cooling or heating costs. Although the lower section exceeds standard room temperature (25°C), the cooling costs would be lower compared to a standard aboveground structures.

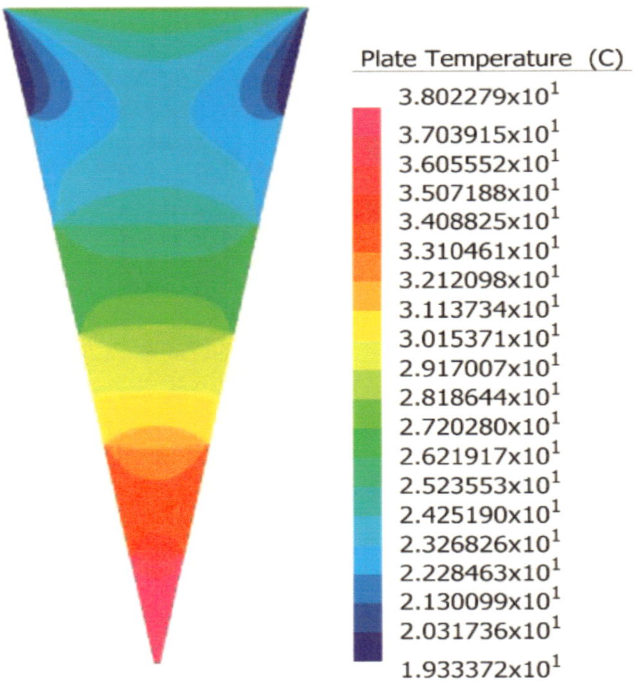

Figure 7.
Temperature contour of steady-state heat solver.

7.2 Earth pressure analysis

7.2.1 Linear static solver

The 3D model shown in **Figure 6** was used to analyze the horizontal displacement of the earthscraper under lateral earth pressures. To achieve this, the model was run under the linear static solver, with the following results produced. **Figure 8** portrays the horizontal displacements under lateral earth pressures. The internal walls experience a maximum deflection of 28 mm, whereas the external walls experience 527 mm of deflection.

7.3 Earthquake analysis

Dynamic analysis can be used to calculate the response under earthquake loading. The acceleration vector, $\ddot{U}$, and displacement vector, U, of a structure under the action of a time-dependent force, F(t), are given by the matrix element equation:

$$\boldsymbol{M}\ddot{\boldsymbol{U}} + \boldsymbol{K}\boldsymbol{U} = \boldsymbol{F}(t), \tag{3}$$

where M is the mass matrix and K is the stiffness matrix of the structure in Eq. (3). This can be solved using the transient solver; however, it is quite computationally demanding. Alternatively, the Strand7's spectral response solver coupled with the natural frequency solver can be used to determine the structure response under an external loading.

7.3.1 Natural frequency solver

The natural frequencies of the structure can be found by removing the external and damping forces and introducing an oscillatory function, U(t), shown below in Eq. (4). This yields the characteristic equation showed in Eq. (5):

$$\boldsymbol{U}(t) = \boldsymbol{U}_0 e^{jwt}, \tag{4}$$

$$\left|\boldsymbol{K} - w^2\boldsymbol{M}\right| = 0, \tag{5}$$

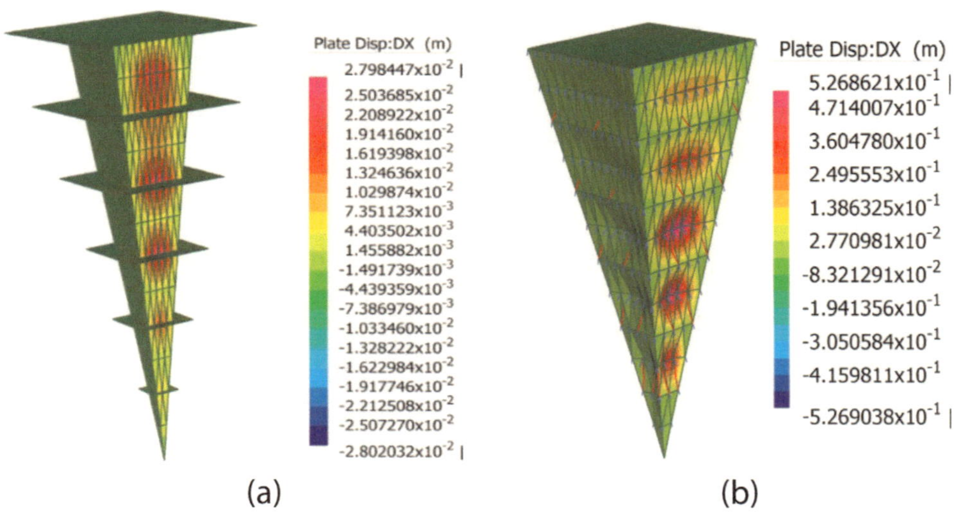

Figure 8.
(a) Interior deflection contour and (b) exterior deflection contour.

Mode	Frequency (Hz)	Mode	Frequency (Hz)	Mode	Frequency (Hz)	Mode	Frequency (Hz)	Mode	Frequency (Hz)
1	0.0933	6	0.5729	11	0.7063	16	0.7847	21	0.8881
2	0.0983	7	0.5829	12	0.7144	17	0.7847	22	0.8882
3	0.0986	8	0.5829	13	0.7144	18	0.8305	23	0.9295
4	0.2818	9	0.5829	14	0.7342	19	0.8312	24	0.9295
5	0.5729	10	0.5830	15	0.7473	20	0.8673	25	0.9960

Table 2.
Frequency and modes of the natural frequency solver.

where w is the natural frequency and the solution gives a polynomial function of w^2 with order equal to the degrees of freedom of the structure. The number of degrees of freedom equals to the number of modes, and the modal shapes are found by normalizing the vibrational modes, shown below in Eq. (6):

$$U_i{}^T M U_i = 1, \tag{6}$$

The natural frequency solver uses this method to find the modal shapes and the respective frequency, w_i, with the modes and respective frequencies found tabulated below. This is as specified by Clause 7.4, with 25 modes found to have converged. **Table 2** shows the different modes and their respective frequencies.

7.3.2 Spectral response solver

The method of spectral response involves determining the peak response of the structure due to an applied acceleration, which is given by the acceleration response shown in Eq. (7):

$$S_a(w_i) = \left|\ddot{S}(t)\right|_{max}, \tag{7}$$

$$S_i(t) = \int_0^t g(t)\frac{\sin\left(w(t-\tau)\right)}{w}d\tau, \tag{8}$$

The spectral response acceleration was input into Strand7 in compliance with AS/NZS1170.4 Clause 7.2, in which the horizontal design spectrum, Cd(T), was determined and is plotted in **Figure 9**.

The maximum displacement is then found by method of square root of the sum of the squares (SRSS) satisfying Clause 7.4.3, in which the contribution of each mode is superposed such that

$$U_{max} = \sqrt{\sum_{i=1}^{n_{dof}} \phi_{i,\,\max}{}^2}, \tag{9}$$

$$\phi_{i,\,\max} = \Gamma_i S(w_i), \tag{10}$$

where Γ_i is the mass participation factor, a measure of the contribution of mode i. A detailed explanation is given in Appendix C. The displacement contour was plotted using the SRSS method. It is shown in **Figure 10** that the maximum displacement due to earthquake loads is 19.8 mm, insignificant relative to the scale of the structure. It should be noted, however, that the design standard AS1170.4 [6] used to generate the numerical analysis does not account for liquefaction,

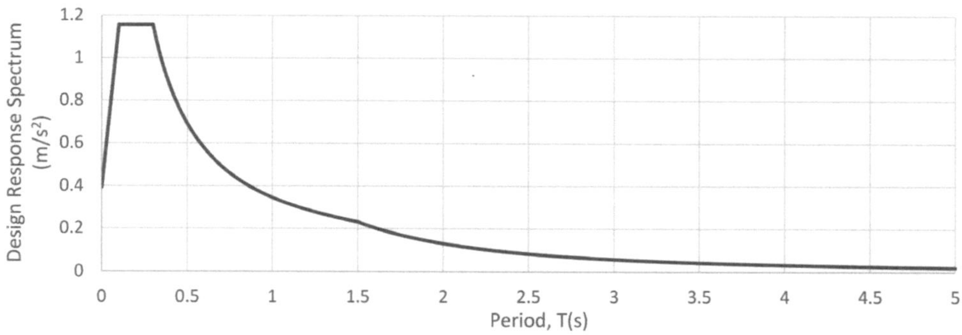

Figure 9.
Design response spectrum.

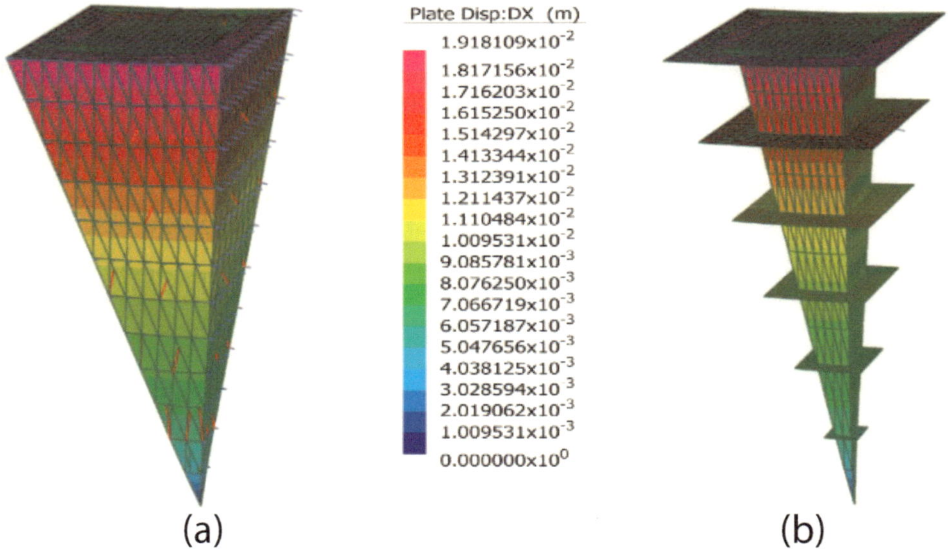

Figure 10.
(a) Exterior deflection under earthquake loads and (b) internal deflection under earthquake loads.

settlement, or fault rupture effects on the soil and as such may be underestimating the effects of earthquake loads on underground structures.

8. Conclusions

This chapter provides a 3D linear static and spectral analysis, as well as a 2D thermal analysis of an inverted pyramid structure modeled off the architectural design of the firm BNKR Arquitectura [8].

Both the 2D and 3D models have been successful in determining the loading conditions of the structure, as well as identifying key structural areas of increased stress, deflection, and heat. Using this analysis, modifications to the structure were made to reduce the effect of these loads and bring the structure within an acceptable level. A final design with a simplified six-floor system arrived upon to allow a sufficiently accurate model while maintaining the viability of 3D modeling within the finite element analysis software Strand7.

While the results found are encouraging, further research should be undertaken to investigate the liquefaction effect of earthquake loads on underground structures and soil. In addition, more detailed structural modeling could be undertaken to investigate the minutia of detail for each individual story, rather than simplifying into six larger general floors.

Therefore, the structure could feasibly be considered in 2050, with a rapidly increasing sophistication of software and material engineering. The demands of the human population in 30 years may necessitate such an ambitious design.

Appendices

A. Earth pressure analysis

Figure 11 portrays the soil pressures experienced by the external walls of the structure, with the relevant values kept parametrized for the determination of Eq. (10).

For substitution in Eq. (10), the parameters and their respective values are shown in **Table 3**. These values were found through literature review and govern the lateral earth pressures experienced by the structure.

The lateral pressure is given as

$$\sigma_h(z) = \sigma_h'(z) + u(z), \tag{11}$$

where u(z) is the pore water pressure, given as the hydrostatic water pressure shown below

$$u(z) = \gamma_w \langle z - d_w \rangle, \tag{12}$$

and $\sigma_h'(z)$ is the effective lateral pressure, expressed as

$$\sigma_h'(z) = K_0 \sigma_v'(z), \tag{13}$$

where K_0 is the coefficient of earth pressure at rest, given by the following expression given by Jaky [7]

$$K_0 = 1 - \sin(\phi), \tag{14}$$

and $\sigma_v'(z)$ is the effective vertical earth pressure. This is the contribution of the different soil layers less the effect of pore water pressure, given as

$$\sigma_v'(z) = \gamma_{ss}(z - \langle z + d_{ss} \rangle) + \gamma_{hs}\langle z - d_{ss} \rangle - \gamma_w \langle z - d_w \rangle, \tag{15}$$

The Macaulay brackets seen above operate as follows:

$$\langle z - a \rangle^n = \begin{cases} 0, & z < a \\ (z - a)^n, & z \geq a \end{cases}, \tag{16}$$

This allows one expression for earth pressure at any depth. Replacing all the relations together gives

$$\sigma_h(z) = K_0(\gamma_{ss}(z - \langle z + d_{ss} \rangle) + \gamma_{hs}\langle z - d_{ss} \rangle) + (1 - K_0)\gamma_w \langle z - d_w \rangle, \tag{17}$$

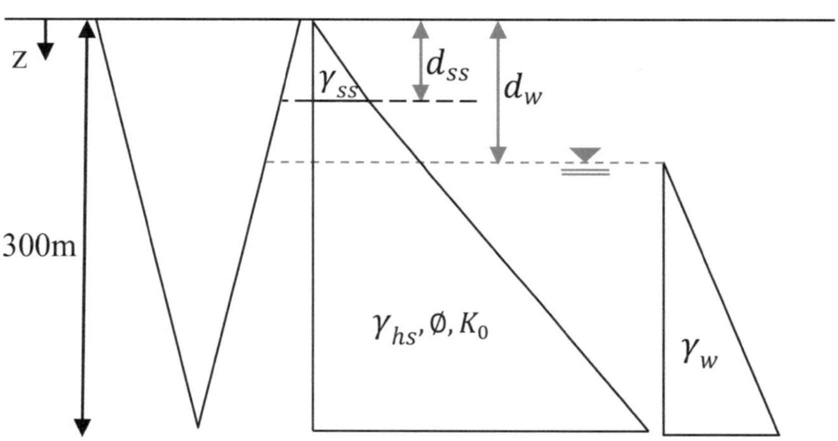

Figure 11.
Earth pressure profile.

Parameter	Value
Internal friction angle, $\phi(°)$	47.2
At-rest coefficient of earth pressure, K_0	0.26627
Soft soil unit weight, γ_{ss} (kN/m^3)	17
Water unit weight, γ_w (kN/m^3)	10
Hawkesbury sandstone unit weight, γ_{hs} (kN/m^3)	24
Depth of soft soil, d_{ss} (m)	10
Depth of water table, d_w (m)	60

Table 3.
Earth pressure parameters.

At the bottom of the structure, i.e., z = 300 m, the lateral earth pressure is calculated as 3.66 MPa, shown below in Eq. (10) using the parameters in **Table 3**:

$$\sigma_h(300) = 0.26627(17(10) + 24(300 - 10)) + (1 - 0.26627)10(300 - 60) = 3.66 \text{ MPa} \tag{18}$$

B. Thermal analysis

The thermal properties of the materials used in the structure are given in **Table 4**. These parameters determine the conductivity and transfer of heat throughout the structure. The reinforced concrete conducts a relatively higher amount of heat compared to the Perlucor ceiling, both absorbing external heat from the soil and surface, respectively.

The heat flux on the Perlucor ceiling can be estimated for a blackbody using Stefan-Boltzmann law. The heat flux generated at the sun's surface, q_{sun}, is given by

$$q_{sun} = \varepsilon\sigma\left(T_{sun}{}^4 - T_{space}{}^4\right), \tag{19}$$

Hence, for radius, R, of the sun and distance D from the sun to the earth, the heat flux experienced by earth is

$$q_{earth} = \varepsilon\sigma\left(T_{sun}{}^4 - T_{space}{}^4\right)\left(\frac{R}{D}\right)^2, \tag{20}$$

The parameters needed for Eq. (17) are shown in **Table 5**, giving the value of

$$q_{earth} = 1 \times 5.67 \times 10^{-8}\left(5780^4 - 3^4\right)\left(\frac{6.957 \times 10^5}{1.5 \times 10^8}\right)^2 = 1.361\frac{\text{kW}}{\text{m}^2}, \tag{21}$$

Material	Reinforced concrete	Perlucor	Air
Coefficient of thermal energy (W/m K)	1.37	0.78	0.0257
Specific heat capacity (J/kg K)	880	840	1005

Table 4.
Material thermal properties.

Parameter	Value
Stefan – Boltzmann constant, σ ($W/m^2\ K^4$)	5.67×10^{-8}
Temperature of Sun, T_{sun} (K)	5780
Temperature of space, T_{space} (K)	3
Distance from sun, D (km)	1.5×10^8
Radius of sun, R (km)	6.957×10^5
Radiation coefficient of sun, ε	1

Table 5.
Heat flux parameters.

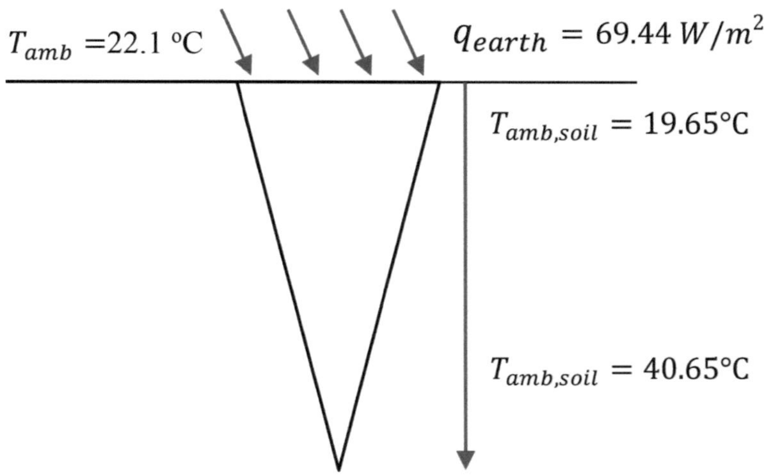

Figure 12.
Thermal conditions.

This however assumes the earth acts as a blackbody, absorbing all incoming radiation. Measurements by the Bureau of Meteorology have shown that the incoming solar radiation is of the value of 69.44 W/m^2, significantly lesser than the blackbody value. This is due to solar irradiance, cloud reflection, atmospheric absorbance, and the reflection of the earth surface. The thermal boundary conditions are shown in **Figure 12.**

C. Earthquake analysis

Pells (2004) determined the bearing capacity of Hawkesbury sandstone as 20–103 MPa and 6–14 MPa for laboratory tests and field tests, respectively. Taking the field tests as true, this classifies the rock as subsoil Class B_e (AS/ZS1170.4 Clause 4.2.2).

Using AS/NZS1170.4, the hazard factor, Z, in Sydney is 0.08. For an important factor of 3 (BCA 2016), the structure has an annual probability of exceedance of 1/1000 (BCA 2016).

AS/NZS1170.4 gives the probability factor of 1.3 for an annual probability of exceedance of 1/1000.

AS/NZS1170.4, for a structure height of 300 m, the earthquake design category is EDCIII. Structural ductility factor (μ) and structural performance factor (S_p) are found as 2 and 0.77, respectively (AS/NZS1170.4).

The spectral shape factor, $C_h(T)$, is found by using AS/NZS1170.4 Clause 6.4. This requires the period of vibration of the structure, which was found using the natural frequency solver. The horizontal design response spectrum, $C_d(T)$, is found using AS/NZS1170.4 Clause 7.2, given as

$$C_d(T) = k_p Z \frac{S_p}{\mu} C_h(T) = 1.3 \times 0.08 \times \frac{0.77}{2} C_h(T) = 0.04004 C_h(T), \tag{22}$$

The scaling factor is multiplied by the gravitational acceleration to give the horizontal acceleration in terms of m/s^2. This process is summarized in **Table 6**, showing the modes, the respective frequency and period, then the calculated and design spectral response factor, and then the mass participation factor. The total mass participation of the 25 nodes is 99.461%, satisfying AS/NZS1170.4 Clause 7.4.2.

Mode	Frequency (Hz)	Period (s)	$C_h(T)$	$C_d(T)$ (g)	$C_d(T)$ (m/s^2)	Γ_i(%)
1	0.0933	10.7209466	0.011484383	0.000459835	0.004510978	0.516
2	0.0983	10.1726597	0.012755718	0.000510739	0.005010349	88.177
3	0.0986	10.1378712	0.012843411	0.00051425	0.005044794	0
4	0.2818	3.5484392	0.104833282	0.004197525	0.041177716	0
5	0.5729	1.7455268	0.433232339	0.017346623	0.17017037	0.001
6	0.5729	1.7455268	0.43323234	0.017346623	0.170170371	0
7	0.5829	1.7156367	0.448459526	0.017956319	0.176151493	0
8	0.5829	1.7155891	0.448484434	0.017957317	0.176161277	0.019
9	0.5829	1.7155890	0.448484477	0.017957318	0.176161294	0
10	0.5830	1.7154045	0.448580981	0.017961182	0.1761992	0
11	0.7063	1.4159108	0.621508079	0.024885183	0.24412365	0
12	0.7144	1.3998054	0.628658824	0.025171499	0.246932408	0.018
13	0.7144	1.3998053	0.628658855	0.025171501	0.24693242	0
14	0.7342	1.3620201	0.646099116	0.025869809	0.253782822	0
15	0.7473	1.3381278	0.657635248	0.026331715	0.258314127	0
16	0.7847	1.2744305	0.690504491	0.0276478	0.271224916	0.007
17	0.7847	1.2744305	0.690504508	0.027647801	0.271224923	0
18	0.8305	1.2041558	0.730802435	0.029261329	0.287053642	0
19	0.8312	1.2031179	0.731432875	0.029286572	0.287301275	0
20	0.8673	1.1530177	0.763214655	0.030559115	0.299784916	0
21	0.8881	1.1259461	0.781564952	0.031293861	0.306992773	10.722
22	0.8882	1.1259172	0.781585013	0.031294664	0.307000653	0
23	0.9295	1.0758671	0.81794486	0.032750512	0.321282524	0
24	0.9295	1.0758671	0.817944861	0.032750512	0.321282525	0
25	0.9960	1.0040134	0.876482347	0.035094353	0.344275605	0

Table 6.
Natural frequency results and spectral response.

Author details

Faham Tahmasebinia[1,2]*, Kevin Yu[1], Jiachen Bao[1], George Chammoun[1], Edwin Chang[1], Samad Sepasgozar[2] and Fernando Alonso Marroquin[1]

1 School of Civil Engineering, The University of Sydney, Sydney, NSW, Australia

2 Faculty of Built Environment, The University of New South Wales, Sydney, NSW, Australia

*Address all correspondence to: faham.tahmasebinia@sydney.edu.au

References

[1] Sol Robles J. La Redención del Ornamento: Recuperando la Dimensión Simbólica de la Arquitectura. RevistArquis. 2012;**1**(2):26-35

[2] McDonald P, Kippen R. Projecting future migration levels: Should rates or numbers by used? People and Place. 2002;**10**:82-83

[3] Morse JM et al. Developing Grounded Theory: The Second Generation. New York: Routledge; 2016

[4] Trahair N, Bradford MA. Behaviour and Design of Steel Structures to AS4100: Australian. New York: CRC Press; 2014

[5] Gilbert R. Creep and shrinkage models for high strength concrete–Proposals for inclusion in AS3600. Australian Journal of Structural Engineering. 2002;**4**(2):95-106

[6] AS AS. Glass in Buildings-Selection and Installation. Standards Association of Australia. 1989

[7] Jaky J. Pressure in silos. In: Proceedings of 2nd ICSM. Japan: ICSM; 1948

[8] Hudson D et al. ACCESS-S1: The new Bureau of Meteorology multi-week to seasonal prediction system. Journal of Southern Hemisphere Earth Systems Science. 2017;**67**(3):132-159

[9] Sharma R, Richards P. The influence of Helmholtz resonance on internal pressures in a low-rise building. Journal of Wind Engineering and Industrial Aerodynamics. 2003;**91**(6):807-828

Chapter 3

Floating Cities Bridge in 2050

Faham Tahmasebinia, Yutaka Tsumura, Baichuan Wang, Yang Wen, Cheng Bao, Samad Sepasgozar and Fernando Alonso-Marroquin

Abstract

A floating cities bridge is designed to connect two floating cities or nearby land to resolve the problem of shortage of construction land due to an increase of population and sea level. The Yumemai floating bridge is referenced as a sample structure; the member sizes and dimensions are modified to suit the need of the project. A finite element structure is built using Strand7, which includes dead load, live load, tidal wave, and wind load. Based on the loads, both static and dynamic analyses are conducted to determine the stress and deflection of the structure. The report outlines the modeling techniques, element types, and analysis solvers used in modeling and analyzing the structure. This report discusses the results obtained from the analysis. The advanced material with low density applied is introduced, which has a good resistance of corrosion and high strength. The main objective of the current chapter is to suggest and design the procedure which can be used as floating structural elements in the future.

Keywords: floating cities bridge, finite element structure

1. Introduction

According to the UN DESA report [1, 2], it is estimated that the world population will reach 9.7 billion in 2050 and 11.2 billion in 2100. This exploding growth of the population brings a great pressure on the demands of construction land. In addition, an increasing sea level aggravates the pressure on the shortage of the land. Sea level will rise as much as 50 cm by 2050 as the report of midrange projections indicated. Floating cities are required to release the intensity of using lands and under such circumstance, and bridges will be an essential infrastructure connecting floating cities. Since floating cities are a mobile structure, the bridge also needs to be mobile and flexible in terms of placement; the solution proposed in the report is a floating bridge which can move around with floating cities and be rapidly deployed when needed. The bridge is designed as modular bridge; thus multiple bridge module can be connected to achieve the desired length depending on the distances required. A floating bridge has a number of advantages over the conventional bridge, for example, marine environment remaining undisturbed; easier construction, as most of construction process can be done onshore; and floating structures being immune to earthquake.

A successful floating bridge, Yumemai Bridge in Japan [1], is referenced as a basic model. Dimensions and sizes are modified to suit the need of the project. The behavior of isolated bridge module is investigated as an initial step to model the bridge. The report presents the Strand7 model, the introduction of the new material, and analysis results associated with the static and dynamic behaviors of the bridge.

2. Structural system

In **Figure 1**, the length of one bridge module is 382 m, width is 38.8 m, and height is 65.5 m, respectively. The number of lanes is designed as four. The pontoon size is 7.5 × 58 × 58 m. The bridge is comprised of a bridge deck which is supported via six columns at each end. The bridge deck between pontoons is hanged by vertical members at 23.3 m intervals. The vertical members are necessitated for this larger-span floating bridge to decrease deflection in the center of the bridge. To control the deflection, the deck itself has a parabolic shape which has a 640 mm rise in the center. The vertical members are supported by the larger-span arch structures. The loads applied to the bridge deck in service are transferred into the vertical member and then passed into the arch structure which carries the load in compression. Finally, the load is carried by the pontoon via the load paths of columns.

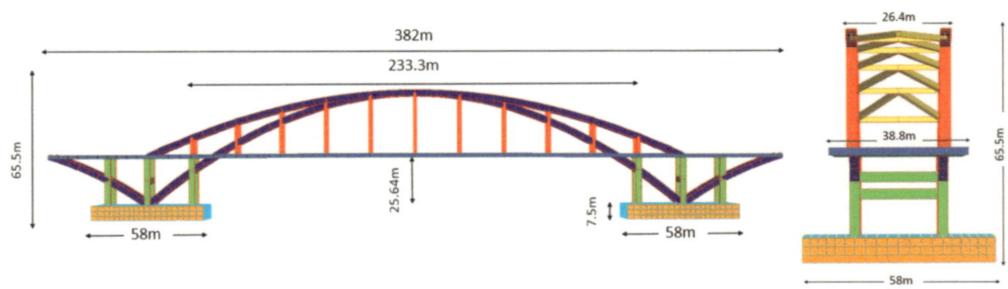

Figure 1.
Dimensions of the bridge.

3. Materials

In a floating offshore structure, the self-weight and corrosion resistance are the two most critical parameters that decide the material to be used in the structure. New novel materials are adopted in this floating bridge, which has characteristics of light, high corrosion resistance, and high strength. These materials enable to reduce the self-weight and increase the durability in aggressive ocean environment. The two main materials used in the floating bridge are nanocrystalline steel and carbon fiber-reinforced plastic (CFRP).

Nanocrystalline steel is chosen as it has an extremely high yield strength. It contains a significant volume fraction of interfacial regions separated by nearly perfect crystals [3]. Since the atomic structure of the grain boundary differs greatly from the inside of the grain, such steel full of grain boundary shows properties that are quite different from those full of conventional metals [4]. The deformation of the metal material is carried out by a sliding motion of lattice defect called as dislocation. In the bulk nanocrystalline metal, the individual crystals are strongly bound by the grain boundaries, and the energy of the dislocation also changes; its behavior surpasses any metals available today. Bulk nanocrystalline steel shows

Material used	Yield strength (MPa)	Yong modulus (GPa)	Density (kg/m^3)
Bulk nanocrystalline steel	2000	200	7850
Concrete (40 MPa)	40	34.5	2400
CFRP	1300	300	16.5

Table 1.
Material properties.

strength which is four times higher than that of conventional metal which is currently developed [4]. It is used for all metal structures of the bridge [3] and is assumed to have a yield strength of 2 GPa which is four times greater than the current structural steel. Such high-strength material allows to reduce the self-weight drastically, by enabling to reduce the cross-sectional area of the member. And also nanocrystalline steel has an excellent corrosion resistance which makes it the ideal material for an offshore floating structure.

Another material used is carbon fiber-reinforced polymer composites (CFRP). It is an extremely strong and light fiber-reinforced plastic which contains carbon fibers. The yield strength of the CFRP in tension and compression are 1600 and 1300 MPa, respectively, Young's modulus 300 GPa, and Poisson' ratio 0.3 with a density of 1.65 g/cm^3 [5]. This material is utilized in the pontoon. Convectional 40 MPa concrete is also used in the columns. The properties of the materials used are summarized in **Table 1**.

4. Loads

4.1 Load analysis

When designing a floating bridge, it is necessary to consider the effects of dead load, live load, wind load, and tidal wave load on the behavior of the bridge. The vertical load resulted from self-weight, and live load is counteracted via buoyancy force. The buoyancy force is generated by using the large pontoons. The live load is caused by traffic, and the truss bridge deck carries the traffic load directly. Wind loads are perpendicular to the face of the members; on the other hand, wave load is applied to the face and side surface of the pontoon, and both wind and tidal wave loads are time-dependent load and hence vary with time.

The bridge deck is supported by the six columns and the vertical cables. And the cable is hanged from two main arch structures. Six columns are supported via large pontoons. The pontoons also support the two main arch structures.

4.2 Calculation of loads

4.2.1 Dead load

Dead load is the self-weight of the floating bridge, since all the used materials' sizes, densities, and other properties are inputted in Strand7; the dead load is calculated automatically by Strand7 under the effect of gravity g = 9.81 m/s^2.

4.2.2 Live load

The floating bridge is catered to connect two floating cities with four lanes for carrying traffic. The estimated traffic loading is determined in according to the AS

5100.2 clause 7.2.4 [6]. The load model M1600 was chosen to determine the traffic load, which is considered as the worst case. This load model means that the bridge is fully loaded with trucks. The uniformly distributed component of the M1600 moving traffic load shall be continuous under the axle group loads and shall be considered as uniformly distributed over the width of a 3.2 m standard design lane.

To make the calculation simple, the total load was applied on the bridge deck as a pressure load. According to the standard, the initial step is to calculate the traffic load for a portion of the bridge and then determine the total load being applied in the whole structure. The load, given for the load model of M1600 from the code, includes the axle group loads of 360 kN and uniformly distributed load of 6 kN/m. There are four lanes of traffic this load multiplied with 4. The total length of the bridge is 382 m, and the bridge is subdivided into 16.4 segments. For each segment, the load with 6360 kN is applied according to the standard. Finally, the total load is 10 kPa in that the load for each segment is multiplied by 15.3.

Length of each segment

$$L = 1.25 \times 8 + 3.75 + 6.25 + 5 = 25\text{m} \tag{1}$$

Number of segment in the signed bridge

$$382 \div 25 = 15.3 \tag{2}$$

Total load on the bridge

$$(360\text{ kN} \times 4 + 6\text{ kN} \times 25) \times 15.3 \times 4 = 97308\text{ kN} \tag{3}$$

Equivalent pressure

$$97308 \div 25 \div 382 = 10\text{ kPa} \tag{4}$$

4.2.3 Wind load

The design wind load speed is determined based on the AS1170 [7], which is 57.8 m/s ARI of 1000-year designed wind speed. The wind load is applied on every member as a pressure load including the deck face that is perpendicular to the wind direction.

Designed wind speed

$$V_{des,\theta} = V_R \times M_d \times (M_{z,cat} \times M_s \times M_t) = 46 \times 1.2564 \times 1 \times 1 \times 1 = 57.8\text{ m/s} \tag{5}$$

Pressure on the side of the deck

$$P = (0.5\rho_{air})[V_{des,\theta}]^2 C_{fig} C_{dyn} \tag{6}$$

$$C_{fig} = C_{dyn} = 1 \tag{7}$$

Side

$$P = \left(0.5 \times \frac{1.2kg}{m^3}\right)\left(\frac{57.8m}{s}\right)^2 = 2.005\text{ kPa} \tag{8}$$

4.2.3.1 Bridge deck uplift and downlift

The wind load applied in the bridge deck is simplified, and the external coefficients ($C_{p,e}$) are −1.3 (up) for top bridge deck and −1.2 (down) for bottom bridge deck which is not varied with position assumedly.

$$C_{fig} = C_{dyn} = -1.3 \tag{9}$$

$$P_{uplift} = \left(0.5 \times 1.2\ \text{kg/m}^3\right)(57.8)^2 \times (-1.3) = -2.606\ \text{kPa} \tag{10}$$

$$P_{downlift} = \left(0.5 \times 1.2\ \text{kg/m}^3\right)(57.8)^2 \times (-1.2) = -2.405\text{kPa} \tag{11}$$

4.2.4 Tidal wave load

The ocean water flows through the pontoon and causes two types of load that are the frontal drag force and the side drag force.

In order to consider worst case, assuming full pontoon is subjected to tidal wave load and this scenario cause the maximum adverse load. According to the AS5100 [6], the load is calculated as below. The velocity of water is assumed typical wave speed under severe storm condition, 14.8 m/s [7] .

Design drag force (in front of pontoon)

$$F_d = 0.5C_dV^2A_d \tag{12}$$

where C_d = *drag efficient* and A_d = *wetted area of the pontoons normal to the water flow*.

$$C_d = 1.4\ (\textit{square shape}) \tag{13}$$

$$A_d = 7.5\text{m} \times 58\text{m} = 435\text{m}^2 \tag{14}$$

$$P = 0.5C_dV^2 = 0.5 \times 1.4 \times (14.83\text{m/s})^2 = 154\text{Pa} \tag{15}$$

$$F_D = 0.5C_dV^2A_s = 154 \times 435\text{m}^2 = 66968\text{N} \tag{16}$$

4.2.5 Side drag on pontoon

$$\text{F}_\text{L} = 0.5\text{C}_\text{s}\text{V}^2\text{A}_\text{L} \tag{17}$$

where C_s = *drag efficient* and A_L = *wetted area of the pontoons normal to the water flow*.

$$C_s = 0.9, \textit{for}\ \theta_w \le 30° \tag{18}$$

$$A_L = 58\text{m} \times 7.5\text{m} = 435\text{m}^2 \tag{19}$$

$$P = 0.5 \times (0.9) \times (14.83\text{m/s})^2 = 100\text{Pa} \tag{20}$$

$$F_L = 0.5C_sV^2A_L = 100 \times 435\text{m}^2 = 43051\text{N} \tag{21}$$

4.2.6 Maximum buoyancy

This floating bridge, compared with normal sea bridge, is dependent on the buoyancy force to support the structure instead of installing pile foundation at the seabed. A pontoon built below the column of the bridge is submerged in the ocean, and it generates a buoyancy force that can balance the opposite vertical load. These vertical loads result from self-weight, wind load, live load, and traffic load. The size of the pontoon is 58 × 58 × 7.5 m; the specific weight of the seawater may be taken to be 10.05 kN/m^3. The maximum buoyancy of the pontoon is calculated as follows:

$$V = 7.5\text{m} \times 58\text{m} \times 58\text{m} = 10765\text{m}^3 \tag{22}$$

$$F_{buoyancy} = \rho g V = 10.9\text{kN/m}^3 \times 10765\text{m}^2 = 108188\text{kN} \tag{23}$$

5. Modeling element and material

5.1 Mesh and mesh quality

The bridge deck is made by plate elements, and all the others are made by using beam elements. All the elements are subdivided to the shape as square as possible; the deck and columns as well as the columns and pontoon have to align with each other to ensure they are entirely connected to each other to get an accurate result and avoid any mesh discontinuity.

5.2 Nodal constraints

The boundary conditions for the left end which are connecting Sydney land, the DX, DZ, RX, and RY are fixed. The other side which is connecting the floating city or next bridge module is free in both translation and rotation. To carry out linear static, natural frequency, and harmonic response analysis, the base of the pontoon is assumed to be fixed in DY, otherwise static condition cannot be established. In nonlinear transient dynamic analysis, the constraints at the base of the pontoon are removed, and pontoon is free in all directions; the constraints are replaced with buoyancy pressure to simulate the dynamic behavior of the bridge on the water.

5.3 Element types

5.3.1 Bridge deck

To comply with AS5100 deflection control and minimize deflection, the deck itself has a parabolic shape with a 640 mm rise in the center. This allows to reduce the deflection under both serviceability and ultimate states. To simplify the model, the truss deck is modeled as a plate element; the truss is assumed to have a relative density of 0.025, and the effective elastic modulus is obtained by the following formula [8]:

$$E_L = \frac{1}{9}\rho_L E_0 \tag{24}$$

where ρ_L is the relative density and E_0 is the elastic modulus of parent material. The pontoon relative density is set as 0.05 and the same procedure is applied.

5.3.2 Main arch

The dimension of the main arch is decided first by solving the simple arch structure applying only live load calculated above; once the maximum force is obtained by simple statics, the required cross-sectional area to carry the applied load is calculated. Outer dimension of the member is based on the referenced structure Yumemai Bridge, and the thickness of the member is calculated by distributing the area required around four sides of the square hollow section member. Then the member thickness is modified to control deflection, and, finally, the main arch was modeled using a beam element with the dimension of 3 × 3 × 0.03 m RHS.

5.3.3 Vertical members

There are 11 vertical members connecting the main arch and the bridge deck on both sides; the height of the members is changing along the bridge with the different positions under the arch; the cross section is 1.5 × 0.75 × 0.015 m. Beam element is used.

5.3.4 Cross member

Between the two main arches, the cross member was applied by using the beam element with the dimension of 0.75 × 0.75 × 0.01 m RHS.

5.3.5 Column

The deck and pontoon were connected by six columns on each end of the bridge; the columns are a 3 × 3 m solid concrete which was modeled using a beam element.

5.3.6 Pontoon

The size of the pontoon is 58 × 58 × 7.5 m cuboid modeled by a solid element in Strand7. Initial submerged depth is set as 4.5 m, i.e., 3 m from the top of the pontoon; the bridge will then adjust itself to gain equilibrium with water pressure, and the final submerged depth will be 4 m.

Dimension and material properties of each element are summarized in **Table 2**.

Modeled element	Modeling shape	FEM element	Material used	Yong modulus (GPa)	Density (kg/m^3)	Geometric properties (m)
Main arch	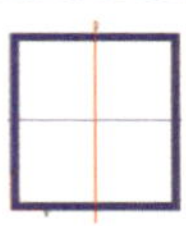	Beam	Bulk nanocrystalline steel	200	7850	3 × 3 × 0.03
Vertical member	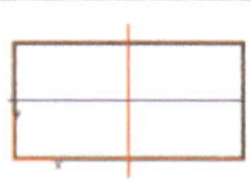	Beam	Bulk nanocrystalline steel	200	7850	1.5 × 0.75 × 0.015
Cross member		Beam	Bulk nanocrystalline steel	200	7850	0.75 × 0.75 × 0.01
Column	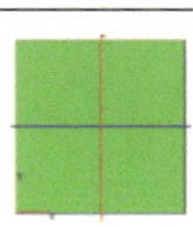	Beam	Concrete (40 MPa)	34.5	2400	3 × 3
Deck		Plate	Bulk nanocrystalline steel (Truss)	0.556	196.25	2 m thickness
Pontoon		Solid	CFRP	1.67	82.5	58 × 58 × 7.5

Table 2.
Element properties.

6. Numerical analysis

6.1 Linear static analysis

The linear static solver is used to analyze the effects of different combination load cases. According to AS1170, the combination load cases are calculated basing on the dead load, live load, and wind load; two critical load cases 1.0G dead load only (serviceability state) and 1.2G + 1.5Q (ultimate state) are considered to check the adequacy of the structure.

By comparing the results for different combination load cases, it is found that load case 2 (ultimate state) has significant effects on the structure. **Figure 2** shows that the maximum displacement under ultimate state is 1600 mm. Note that the bridge deck is raised by 640 mm, so under the ultimate condition, the deflection of the bridge deck is 960 mm that is below the flat level. It is also found that the maximum stress is 254 MPa compression which is located in the root of the main arch structure. In **Figure 3**, under serviceability state (1.0 G), the maximum deflection of the deck is 505 mm which is 95 mm above the flat level. AS5100 [6] limits the deflection under serviceability condition to be "deflection/span < 1/600"; the bridge complies the deflection limit. The maximum stress under serviceability condition is found to be 82.7 MPa.

The linear static analysis showed that the material dimensions are satisfactory, both in terms of strength and deflection, under serviceability and ultimate states.

6.2 Natural frequency analysis

The natural frequency is defined as free vibrations of an elastic body. If the frequency of the applied load is identical as the natural frequency of the structure, hence the amplitude of vibration increases manifold. It is significantly important to find the critical frequency that causes the maximum deflection. Two hundred different natural frequencies under 12 Hz are found of which 173 of them are

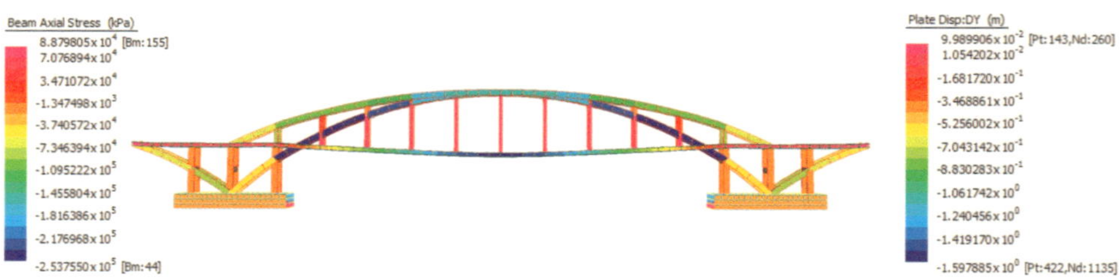

Figure 2.
Stress and deflection under ultimate state.

Figure 3.
Stress and deflection under serviceability state.

converged; the minimum mass participation factor is 85.08%. The typical ocean wave has a frequency ranging from 0.1 to 5 Hz [8]. The analysis covers this area of the frequencies. These frequencies found are used to conduct harmonic response analysis. The first 20 modes of natural frequencies are listed in **Table 3**.

6.3 Harmonic response analysis

In floating structure the effect from the earthquake is minimal as the water acts as a base isolation material; hence any ground movement will not be transferred to the structure, but what matters to the structure is the daily wave and wind load. As mentioned previously, the typical wave has a frequency between 0.1 and 5 Hz. Natural frequency analysis in **Figure 4a** showed that there are a number of natural frequencies within this range; when the wave frequency coincides with the natural frequency, the effect on the structure could be significant and cause devastating damage. Thus it must be investigated. Harmonic response solver is used to test the effect of sinusoidally varying the load under 7.5 Hz; the dominant varying load will be the wave load, however, to be conservative, and wind loads are also assumed to be a varying load.

The deflection at the center of the deck versus different frequencies is plotted in **Figure 4a**. The harmonic response analysis has shown the frequency of 0.51789 Hz produced the largest deflection, and it can be said that the structure is most sensitive to this frequency. **Figure 4b** shows deflection contour under such frequency; the maximum displacement due to the frequency is found to be 0.456 m since the deflection caused by the load is relatively small considering the large span of the bridge, and it also complies the 1/600 deflection limit. The bridge will still be able to serve its purpose.

6.4 Nonlinear transient dynamic analysis

The nonlinear transient dynamic solver is used to analyze the effects of the combination dynamic loads on the floating bridge. Under transient dynamic circumstance, the effects of the wind load, wave load, and buoyancy force have a significant impact on the bridge deflection and the maximum stress within the structure. Both wind load and wave load are varied with time and utilizing load

Mode	Natural frequency (Hz)	Mode	Natural frequency (Hz)
1	0.0461395	11	1.83116
2	0.349325	12	1.84217
3	0.517894	13	1.96018
4	0.778749	14	2.06939
5	0.929989	15	2.20666
6	1.01908	16	2.29274
7	1.16268	17	2.36013
8	1.21482	18	2.44297
9	1.46656	19	2.52935
10	1.51455	20	2.61975

Table 3.
Natural frequency of the bridge.

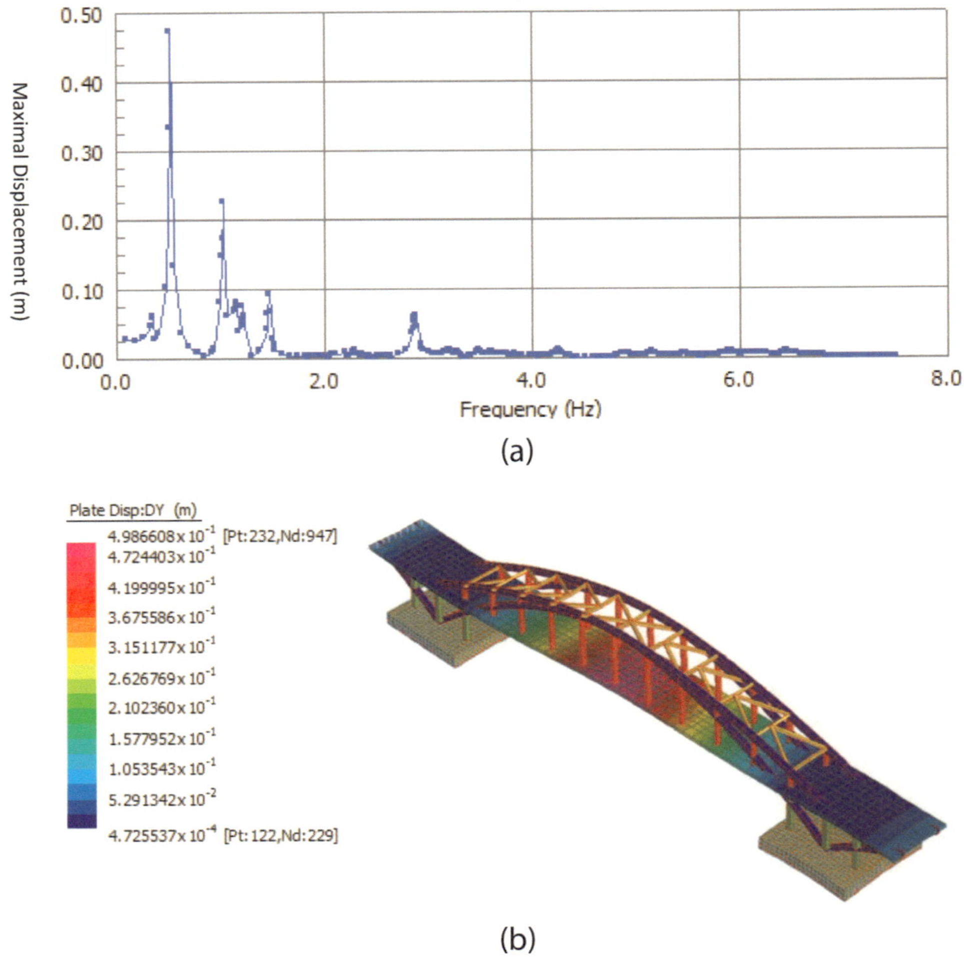

Figure 4
(a) Displacement versus frequency. (b) Displacement contour under varying load f = 0.51789 Hz.

factor variation. The buoyancy force is varied depending on the vertical position of the bridge.

The applied loads, wave load, wind load, and live load, vary with respect to time in nonlinear transient dynamic analysis, and their variation is shown in **Figure 5a–d**. Variation is expressed as a load factor from 0 to 1, one meaning maximum designed load. Tidal wave load is considered as a simple harmonic motion. The designed wave velocity is the typical velocity (14.83 m/s) under storm condition, period of 14.3 s (0.699 Hz), wavelength of 212.6 m, and amplitude of 7.4 m [5]. The wind speed variation is considered as a storm day wind profile in Sydney harbor on October 05, 2017 [8], and shrunk into 140 s as shown in **Figure 5c**; load factor is the ratio between designed wind speed and applied wind speed. One means applied wind speed = designed wind speed (i.e., 58 m/s). The traffic load also varies the simulating traffic condition on the bridge as shown in **Figure 5d**. The trapezoidal loading curve simulates the accumulation of traffic caused by red traffic light and clearing off when the light turns green.

Normal pressure is applied at the bottom of the pontoon to simulate the buoyancy. One kilopascal is applied and factored by its vertical position. Load factor is shown below.

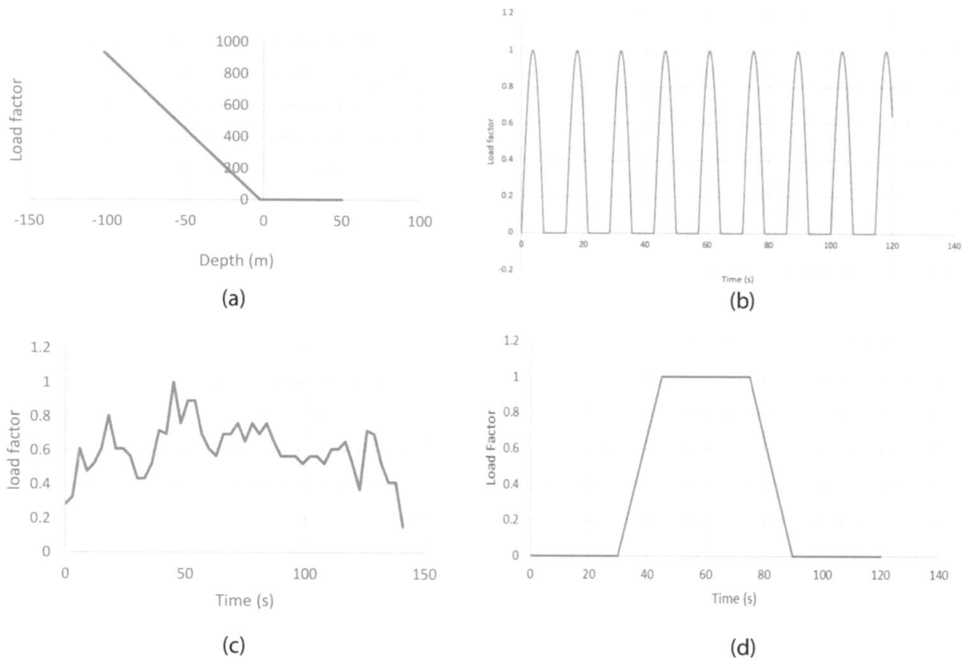

Figure 5
(a) Load factor versus position (buoyancy). (b) Load factor versus time (wave load). (c) Load factor versus time (Wind load)—storm day wind profile in Sydney harbor on 10/05/2017. (d) Load factor versus time (Traffic load)—simulates the accumulation of traffic and clearing off.

7. Results

The behavior of the bridge under dynamic condition is modeled for 2 minutes. The bridge must satisfy the following criteria for it to be successfully designed. Firstly, the maximum stress on the member does not exceed the member capacity. Secondly, the buoyancy provided by the pontoon is able to carry the maximum designed load. And finally, the deflection under serviceability condition complies AS5100 [6].

7.1 Pontoon displacement

As shown in **Figure 6**, the initial oscillation from 0 to 30 s is due to bridge adjusting itself to an equilibrium position; after 30 s the pontoon submerged due to increasing live load and stays at a constant around −2.2 m. At 75 s the pontoon starts float back up as live load is decreased, and about 90 s the bridge is at equilibrium with its self-weight; the small oscillation observed from 90 to 120 s is due to the applied wave load and wind load. As the water level is set at −3 m and the top of the pontoon is 0 m. The maximum displacement of the pontoon can be found at 45 s with a displacement of −2.4 m; at this point the top of the pontoon is 0.6 m above the water level. The bridge is in equilibrium when the pontoon is submerged 4.5 m in the water, and there is 3 m of extra buoyancy, for live, wind, and wave loads, which is equivalent to 200,000 kN of force. When fully designed live load is applied, there is still 0.6 m of extra buoyancy which is equivalent to 40,000 kN, and therefore the size of the pontoon is satisfactory providing sufficient buoyancy under all condition.

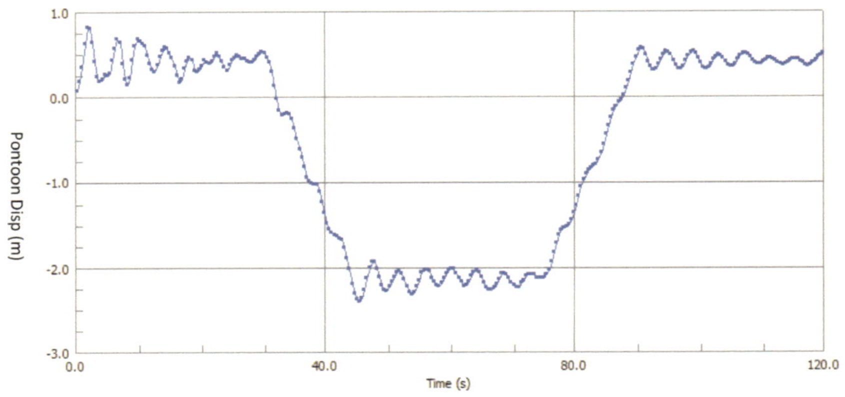

Figure 6.
Displacement at the top of pontoon.

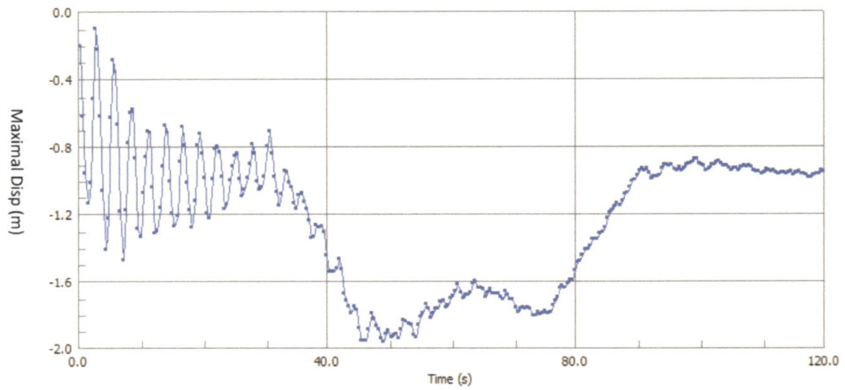

Figure 7.
Deflection at the mid-span of the deck versus time.

7.2 Deck deflection

By comparing the relative displacement of the pontoon and the deck, the deflection due to self-weight, wave load, and wind load can be isolated.

As shown in **Figure** 7 (90–120 s), without traffic load, the deflection of the plate is around 900 mm. Given that the deck has risen by 640 mm in the mid-span, the deck is 260 mm lower than the flat level, and it complies with serviceability deflection of AS5100 [6]. Hence the deflection is also satisfactory. It can be seen that when designed traffic load is applied (45–75 s), the deflection of the plate will increase to 2 m meaning the deck is 1360 mm lower than the flat level. This is greater than the deflection limit of AS5100 [6]. However such case is regarded as an ultimate state; hence the 1/600 deflection limit does not apply.

7.3 Maximum stress

The maximum axial force is found at the root of the main arch structure. Axial force for the member is shown in **Figure 8**. It is observed that the axial force changes with time. The maximal tension (positive) is approximately 5000 kN, and the maximal compression force (negative) is 80,000 kN. The maximum stress can be calculated by dividing the cross-sectional area of the main arch, and hence the maximum stress is 225 MPa.

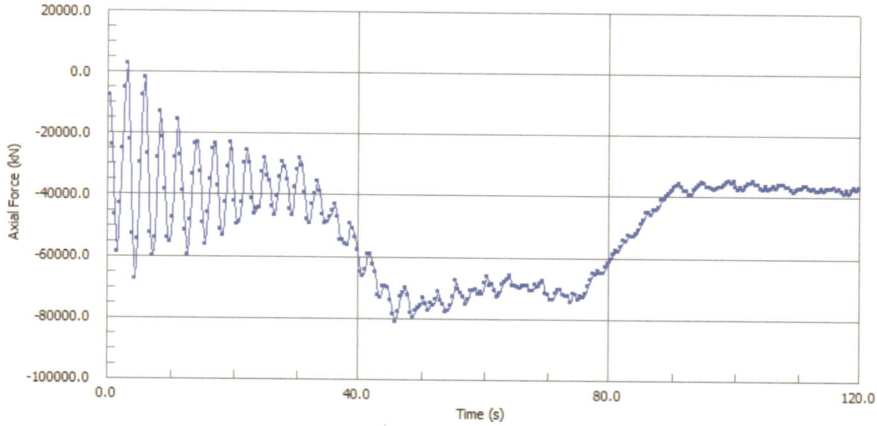

Figure 8.
Axial force versus time.

Loading condition	Maximum stress (MPa)	Maximum deflection (mm)
Linear static analysis (1.2G + 1.5Q)	254	960
Nonlinear transient dynamic (45–75 s)	225	1360
Linear static analysis (1.0G)	82.7	−100
Nonlinear transient dynamic (90–120 s)	112	260

Table 4.
Maximum stress and deflection under different loading condition.

The applied dynamic loads include the traffic load, wind load, and tidal wave load which are 10, 2.5, and 0.15 kPa, respectively. The dominant load is the live load. In the figures related to load factor, it is observed that both wind load and tidal wave load are consistently being applied to the structure. After 30 s, the live load is increasingly imposed to the structure. The figure of displacement and stress can show the effects of the load.

Initially the deflection and stress are caused by the movement of the bridge adjusting the position. Then, the live load starts to influence on the bridge after 30 s. Since the traffic load is significantly larger than other dynamic load, the effect on the displacement and stress is dominated by the traffic load. It is observed that the displacement and stress stay almost constant when the traffic load factor becomes 1 (i.e., when fully designed traffic load is applied) and kept constant from 45 to 75 s. After the traffic load is fully cleared at 90 s, the displacement and stress observed are solely due to wind load and tidal wave load. The stress and displacement fluctuate at the period of the wave load. However, the variation is fairly limited as the bridge contains a large mass and the effect from projected area pontoon is small. Comparison between worst-case static analysis and dynamic analysis is shown in **Table 4**.

8. Discussion

It is interesting to note that in worst-case scenario in linear static analysis which is ultimate load case of 1.2G + 1.5Q produced largest stress in the main arch with 254 MPa with 960 mm of deflection at the mid-span of the deck. On the other hand, in dynamic analysis, the maximum stress is found to be 225 MPa which is lower than the static analysis, and yet the deflection is found to be 1360 mm which is much greater than the linear static analysis. This larger deflection can be explained by the superposition effect that is the sum of response of each stimulus individually. When the peak of wind/wave load coincide with the each other, and since wave load is modeled as sine function with frequency of 0.699 Hz and as natural frequency solver has shown that the one of the natural frequency of the bridge is 0.69798 Hz and is very close to the frequency of the applied wave (0.699 Hz). As a result, the bridge will resonate with the wave loads and produce a larger deflection. This is also evident in the maximum stress in the member; the maximum on the member is fundamentally the function of the load applied; in ultimate state the total applied load 1.2G + 1.5Q produces a greater total load which results in a higher stress in the member. This means the large deflection is not caused by applied load but rather oscillation of the bridge, resonating the wave load.

The maximum stress presented in a member is 254 MPa; the conventional structural steel will be sufficient under such load. However in an aggressive marine environment, the corrosion of the member will be a large issue, and therefore nanocrystalline steel is still the preferred option. However knowing that the stress is in such magnitude, nanocrystalline aluminum will also be an ideal candidate as it is three times lighter than steel. And by the utilizing nanocrystallization, the low yield stress of conventional aluminum could be overcome, and also aluminums are inheritably a more resilient corrosion than steel. There is a drawback of using aluminum such as them having lower elastic modulus; since nanocrystallization does not influence the elastic modulus of the material, the deflection could be a significant issue. Hence size and dimension of the member need to be modified to compensate for the lower elastic modulus.

9. Conclusions

The floating bridges are a practical infrastructure connecting floating cities, which resolve the potential issue of the land shortage due to an increasing population and sea level. The pontoon is utilized to support the structure instead of pile foundation being fixed in the seabed which greatly contributes mobility and thereby feasibility of the floating cities. The primary concept of the designed bridge is "flexibility." The modular design allows easy adjustment to meet specific local demands of a wider range of situation, which renders the structure a suitable solution to the floating cities bridge.

The investigation on effects of static loading and dynamic loading (wind, wave, and traffic loads) on the structure was performed. And it was found that dynamic load produced the largest deflection due to superposition effects and the static loads produced the largest stress. In either case, the design of the bridge fulfills the requirements of deflection, pontoon displacement, and maximum stress. This floating bridge can be referenced for future real project.

In addition, the software analysis cannot conduct the analysis of the influence of the turbulence caused by fluid flowing through the bridge, including wind and water. Computational fluid dynamic (CFD) modeling can be done, and/or a prototype of a floating bridge is necessarily built to simulate a real scenario of wind and ocean wave. Also, the marine environment provides many uncertainties affecting the durability of the bridge. For instance, the salty water and air accelerate the corrosion of the structure. More research should be conducted in the next step.

Author details

Faham Tahmasebinia[1,2]*, Yutaka Tsumura[1], Baichuan Wang[1], Yang Wen[1], Cheng Bao[1], Samad Sepasgozar[2] and Fernando Alonso-Marroquin[1]

1 School of Civil Engineering, The University of Sydney, Sydney, NSW, Australia

2 Faculty of Built Environment, The University of New South Wales, Sydney, NSW, Australia

*Address all correspondence to: faham.tahmasebinia@sydney.edu.au

References

[1] Watanabe E et al. Yumemai floating swing arch bridge of Osaka, Japan. In: Large Floating Structures. Singapore: Springer; 2015. pp. 61-90

[2] You D et al. Global, regional, and national levels and trends in under-5 mortality between 1990 and 2015, with scenario-based projections to 2030: A systematic analysis by the UN Inter-agency Group for Child Mortality Estimation. The Lancet. 2015; **386**(10010):2275-2286

[3] Bhadeshia H. Bulk nanocrystalline steel. Ironmaking & Steelmaking. 2005; **32**(5):405-410

[4] Tsuji N. Unique mechanical properties of nanostructured metals. Journal of Nanoscience and Nanotechnology. 2007;7(11):3765-3770

[5] Pham QP, Sharma U, Mikos AG. Electrospinning of polymeric nanofibers for tissue engineering applications: A review. Tissue Engineering. 2006;**12**(5): 1197-1211

[6] Standard Australia, AS5100.2-2004. Bridge design—Part 2: design loads. Sydney, Australia: Standards Australia; 2004

[7] Standards Australia. SAA Loading Code Part 2: Wind Loading. Australia: Standards Australia; 1989

[8] Zhang Q et al. A study of the effective elastic modulus of a lattice truss panel structure by experimental and theoretical analysis. Composite Structures. 2017;**165**:130-137

Chapter 4

A GIS-Based Risk and Safety Analysis of Entrance Areas in Educational Buildings Based on Students' Experience

Sara Shirowzhan, Laurence Kimmel, Mohammad Mojtahedi, Samad Sepasgozar and Jack Peacock

Abstract

The architecture of educational buildings is required to meet the contemporary needs and follow regulations concerning safety issues in an intelligent, resilient, and systematic manner. However, the current literature generally neglects to identify geo-referenced risks within GIS environment based on the users' perceptions of educational environments. This research aims to present a newly formulated risk evaluation criteria for assessing the spaces within and outside educational buildings. This chapter investigates students' experience of using different micro-spaces, architects' predesign assumptions and expectations of the spaces, and post-design assessment of the spaces. Two case studies of educational spaces within University of New South Wales campus in Kensington, Sydney was selected, and the questionnaire method was employed to collect data from students, who routinely use the selected areas. By comparing the results of the two buildings and mapping them in GIS, it is suggested that feelings of safety and security can be increased via improving the building features and enhancing the building control and security control, for example, installing CCTV and other security infrastructure. Data-driven findings mapped into GIS create a prototype for the identification of problematic areas on a map. The results help decision-makers to understand risks and strengthen risk reduction strategies. This work is also a step towards smarter buildings and enhanced preparedness for an effective response to a security threat, both minimal and extreme.

Keywords: smart cities, resilient cities, safe buildings, public building, GIS

1. Introduction

Crime prevention through environmental design (CPTED) is an established design methodology that aims to reduce the opportunity for crime, fear, and disorder within the built environment. Pioneered by US architect Randall Atlas, this method for designing recognises the intended use of a space in a building and the opportunity to use design elements to achieve security goals without the imposition of traditional security mechanisms (e.g. alarms, gates, and locks). It is important to note that this method can be applied not only to new design projects but also retroactively for existing sites [1]. The CPTED establishes a framework for assessing a program or building, aiming to consider and classify assets, threats, vulnerabilities, and countermeasures, working to establish the security needs and requirements of a site. Through this site assessment, architects and designers should be able to follow guidelines that ensure the safety of clients and users from a range of security threats, from basic petty crime to more extreme terrorist attacks [1].

However, this design methodology is a general guideline and does not provide specific, analysed case studies nor those showing a mapping of how crime and unsafe areas are linked to user perception. This study aims to provide a methodology for mapping an area or location and addresses questions of perceptions of safety in a public, educational environment. The first question aims to identify whether a student feels safer in an outside public space in front of the building or inside the building. What are the factors that would make one access feel safer than another? Do students feel safer if they are more visible by other students and by staff members (directly or indirectly through closed-circuit TV or CCTV cameras)? Outside the building, it is assumed that students have clear vision of the surrounding areas and have the possibility to take different paths if they notice something unusual in the area. The interior of the building could be perceived as more protected, but it is more constraining and generally more difficult to locate a safe space when faced with a threat inside a largely open-plan public building.

Within the interior of the ground floor, what are the factors that would make one location feel safer than another? Considering the factors that might influence the response of students (e.g. gender, time of day or night studying, cultural background, familiarity with area), is it possible to have a map that represents the feeling of safety of the community of users of a building? If yes, what is the accuracy of this map? This study will provide guidelines to help designers and stakeholders in evaluating the perception of safety of the users of their public buildings and also assist them to identify areas in a public building where the perception of safety is low, in order to enhance the well-being of the users.

2. Literature review

This section reviews relevant papers in three main areas to develop an interdisciplinary framework to be a base for conducting a systematic survey. Three main areas are selected, and relevant papers are reviewed as follows.

2.1 Resiliency in public buildings and international frameworks

Safety and security in public buildings is a considerable phenomenon that has been widely debated. The aim is to anticipate hazards in order to alter the frequency and intensity of hazards and disasters within public buildings. This can affect considerably built-up, urban environments, particularly public buildings such as university campuses (Basher, 2006; Tozier de la Poterie and Baudoin, 2015). Disaster risk management studies have focused on natural disaster risks such as fire, floods, and storms, when man-made disasters such as a terrorist attack have been neglected. Using a review of existing literature, this study explores international frameworks for disaster risk reduction (DRR) and resiliency enhancement in public buildings. Methods exist to significantly reduce the number of disaster risks and loss of life, livelihood, and health in the economic, physical, social, cultural, and environmental assets of persons, businesses, communities, and countries [2]. International frameworks for disaster risk reduction and resiliency have very ambitious targets to reduce global hazard incidents in terms of the number of people affected globally and direct disaster economic loss in Gross Domestic Product and the number of countries with high national and local disaster risk [2]. Hence, this research will review disaster risk reduction and building resiliency associated with safety and security within the Sendai Framework for Disaster Risk Reduction 2015–2030 (SFDRR), analysing how public buildings such as university campuses are addressed.

In Australia, local councils are responsible for approving the development of most projects and building plans [2–4]. They are important for reducing risk in planning and building; they must pass bylaws on building regulations and prepare land use plans and emergency response plans. This research seeks to measure public buildings' approaches and synergies with the Sendai Framework for Disaster Risk Reduction 2015–2030. Local councils have emphasised the need to focus on urban resilience; however, there have been insufficient tools and techniques to measure public buildings' resilience efforts with clear quantitative decision-making tools [5].

There are four main priorities explained in the Sendai Framework. These following priorities are split into national/local levels and global/regional levels. It enables effective macro- and micromanagement to tailor towards different areas:

Priority 1: Understanding disaster risk.

Priority 2: Strengthening disaster risk governance to manage disaster risk.

Priority 3: Investing in disaster risk reduction for resilience.

Priority 4: Enhancing disaster preparedness for effective response and to "Build Back Better" in recovery, rehabilitation, and reconstruction.

These priorities of the Sendai Framework for Disaster Risk Reduction are the background framework of this chapter investigating a connection between safety in building and resilience.

2.2 Safety-oriented design

Resilience in the case of a university building means that the design is keeping users and assets safe, and at the same time, users are not anxious in their everyday life by the safety-oriented design. The university building stays open and welcoming to staff, students, and temporary guests. Australia's current counterterrorism guidelines focus on securing crowded places, and universities are considered in this category, especially during special events like open days.[1] While the application of the guidelines to a particular site is informed by criteria set out in [6], the strategies adopted in specific cases—including the extent to which "openness-friendly" measures are adopted—rely on relevant design stakeholder engagement. The outcome of the literature review on safety-oriented design in Australia, the United Kingdom, and the United States[2] is that two broad approaches to securing public spaces can be identified: *enclosing* and *layering*.

Enclosing, also known as "fortressing" (or "traditional target hardening"), is a defensive strategy that focusses predominantly on denying access to a target through exclusion. In its most elementary form, enclosing secures public domains via physical or artificial barrier techniques such as bollards and security walls ([1]: 5, 13).

---

[1] As of 29 June 2019, Australia's ANZCTC has published 16 counterterrorism related guidelines and reports: https://www.nationalsecurity.gov.au/Media-and-publications/Publications/Pages/default.aspx

[2] As of 29 June 2019:

- 13 counterterrorism related guidelines and reports have been published in the US (FEMA website). Those of most relevance to the present discussion are those published by Federal Emergency Management Agency (FEMA): FEMA 426 Reference Manual to Mitigate Potential Terrorist Attacks Against Buildings [7] and FEMA 430 Site and Urban Design for Security [8].
- 25 counterterrorism related guidelines and reports have been published in the UK (gov.uk website). The CPNI (Centre for the Protection of National Infrastructure) issues guidelines for important and sensitive assets (CPNI website). Those of most relevance to the present discussion are the guidelines issued by HM Government [9, 10] and by the Royal Institute of British Architects [11].

Typically, the implementation of this design strategy sees significant standoff distances between entrance points and the building, facades constructed of hardened materials, and the building's interior protected by both secure outer and inner perimeters ([12]: 2.23).[3]

Enclosing has proved a common strategy under the US counterterrorism guidelines for many years, notably in the safeguarding of schools and college campuses.[4] While its implementation offers a very direct (and efficient) enhancement of safety, in design terms it poses significant challenges to maintaining physical and psychological openness. As Randall I. Atlas notes, "[c]onsider the 1970s fortressing in schools and the windowless and sterile physical structures that emerged. That fortressing sent a message of paranoia and fear" ([1], 283).

In contrast to enclosing, layering is a security strategy that focusses on protecting a site by introducing multiple layers of complementary protective measures. The strategy capitalises on redundancy in a site's security design: if implemented correctly, it ensures that the failure of any single layer—which may consist of different security measures—will not significantly compromise the overall security of the building. Rather than reducing the likelihood of a security threat, the goal of layered security is to reduce the likelihood of a *successful* threat. Also known as "security in depth" ([15], 14), strategies of layering consider a building's security both in its proximate context and in its internal layout. Under the US counterterrorism guidelines, FEMA outlines a layered security comprising three layers:

- "First Layer of Defence": comprises barriers (commonly at a property line or sidewalk/curb line) ([7], 2–12). In the context of HV mitigation in front of a building, this first layer may take the form of permanent or removable (but well-designed and integrated) bollards.

- "Second Layer of Defence": extends from the perimeter of the site to the exterior face of a building ([7], 2–12). "The most basic elements of architecture are themselves part of the security systems for buildings: walls, doors" ([16], 16).

- "Third Layer of Defence": commonly comprises the building's interior and separates unsecured from secured areas ([7], 2–12).

More nuanced a security strategy than the defensive "shield" of enclosing, layering—in principle—aligns more naturally with objectives of openness in public building security and thus with objective of resilient design.

2.3 Mapping of perception of safety

Existing research mainly focuses on crime mapping and safe zone mapping in public spaces. Crime mapping is one of the main purposes of recent crime analysis. However, the use of maps for criminology relates to predigital and satellite mapping techniques, when maps were 2D and containing less information.

Since the early 1980s, there has been a resurgence in the interest for three-dimensional analysis of unsafe places and crime areas in order to identify high-risk areas and to visualise information on unsafe areas [17, 18]. When geographical information systems (GIS) were developed in the 1990s, policy makers, including law enforcement departments, started learning more about their capabilities. Policy

[3] Among others, the enclosing strategy has encouraged forms of perimeter protection such as 'Forced-Entry-Ballistic Resistant' (FE-BR): see Whole Building Design Guide [13].

[4] FEMA has published guidelines specifically in relation to school security [14].

makers started learning the basics of location analysis and spatial information in order to identify and develop methodologies to prevent and protect public areas from crime.

The recent literature recommends the use of location-based systems and GIS to detect high-risk areas and the possibility of quick responses to the crimes [19, 20].

Emerging technologies such as advanced sensors, apps, and advanced GIS features are helpful for safety and crime analysis. Different layers of GIS can show the historical or current high-level crime areas and unsafe locations. Relevant data seeks to identify spatial patterns related to space, people activities, location features, and time. However, the utilisation of such technologies now updated [21, 22] and the upscaling to advanced technologies is slow. Several unanswered questions are mentioned in the literature and are knowledge gaps such as [17, 21, 22] what methods such as univariate methods can be used to extrapolate location-based patterns in small public areas. Crime and safety location analysis and mapping demonstrate, measure, analyse risks, and evaluate or rate locations. The analysis also helps to interpret what should be done in high-risk areas, what should be the priorities, and what should be improved. Locating, mapping, and geo-referencing may reveal patterns that can be used to model space and time in predicted future situations [17, 20].

The development and utilisation of GIS for crime and safety analysis is a slow process. The literature gives a couple of reasons for such a slow process, such as the costs, hardware, and software available to public place managers, and the complexity of the system development process [17, 23, 24]. However, it is known as a powerful tool to support crime and unsafe place analysis including the day-to-day activities of place users [23]. At the technical level, GIS uses vector data representing the fundamental units of spatial data such as points, lines, and polygons. The point is a discrete position similar to a pin or flag on a 2D drawing. The line also represents more than two points on a map which is the same as a normal line or a set of points in one line representing the boundaries and walls. The complicated feature can be a polygon representing a geographic area in the selected public building with a set of line segments [25]. Buildings, parks, or lakes are just a few examples. The GIS environment includes a series of elements corresponding to a data bit or a record which should be previously defined as an attribute. **Figure 1** shows the proposed workflow to select the area, collect data, and populate the geodatabase for the safety and crime analysis in this paper.

Since the safety and crime analysis is a complicated analysis, the vector data is not enough for modelling and accurate interpretation. Therefore, the second type of data is required, namely, raster data model. Raster data is RGB or pixel-based data

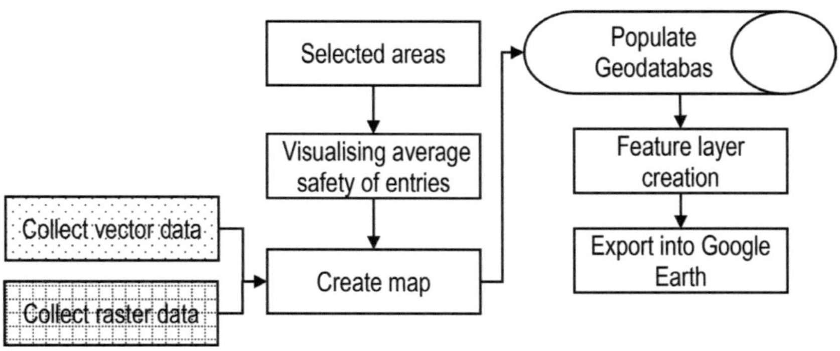

Figure 1.
The proposed workflow for creating and updating the maps in GIS environment.

such as an image. This type of data can be collected from different sources such as aerial or satellite photos. Similar to the vector data, the data record has an attribute; it means that each pixel linked to an attribute value. The combination of both vector and raster data sets will enrich the designed GIS model by including the qualitative data related to geographical data.

Since the public space is selected in an educational area, the literature in the education is also reviewed. The student feelings and attitudes that are obtained by a university's environment can be called the university climate in terms of safety (Loukas, 2007). This is related to the concept of safety and safe learning environment. The quality of the physical environment and a safe university helps students to flourish socially and emotionally.

Previous studies focused on several factors influencing the safety climate in an education place (Loukas, 2007). However, details of the physical environment have been largely ignored. Furthermore, different zones in a large university have different attributes, and the student's feelings are not the same. A question has remained on what main factors potentially influence the university climate. Particularly, what characteristics may contribute more in making feeling higher safety in buildings by students (**Table 1**).

Previous studies explained that the quality of the educational building and infrastructure may affect the student achievement; additionally, this also could affect both students and teachers' attitude towards learning and teaching, respectively. Many papers investigate the relationships between various building design types and features and students' outcomes (Earthman, 2002, 2004; Higgins et al., 2005; Lemasters, 1997; Schneider, 2002, Buckley et al., 2004). For example, Buckley et al. (2004) suggests that the buildings' shape affects the learning outcomes. As another example, Uline and Tschannen-Moran (2008) find that building age, climate control, indoor air quality, lighting, acoustical control, design classifications, and overall impression are key factors. While these studies make relationships between the level of safety and students achievements, they rarely provide detailed information of perceived safety from students' points of view in the corners and other interior spaces of the educational buildings. Also, mapping the perceived safety for better identification of problematic areas within universities has remained scarce.

ID	Variable	References
SAF	The level of safety in my study place	[26, 27]
PVC	The level of privacy in my study space	[28, 29]
SPS	Feel as safe in an enclosed more private study place	[30, 26]
SRV	You feel if your workspace wasn't monitored by surveillance cameras	[31, 32]
BUR	Bollards potentially lowering the threat of unsafe reactions	[33]
VAS	Access influence, e.g. visibility, accessibility, or safety reasons	[34]
COA	affect your choice of access/circulation point	[35]
BLA	Building layout and architecture	[36]
CRA	Card readers/swipe card access	[37, 38]
ASO	Awareness and availability of the security office who monitor the security camera footage	[39]
AWH	Awareness and availability of help points	[40]

Table 1.
A summary of key variables for conceptualization, Shirowzhan, S 2018.

3. Research methodology

This paper is based on an empirical study on two selected educational buildings. A series of semi-structured questionnaires were designed based on the literature. The questionnaires assisted the researchers to translate the information given by the participants into maps and to evaluate the spaces at two relative local or global levels considering other spaces around the selected areas. A group of participants from each building was interviewed, and the results analysed. A comparison method was chosen to compare two selected educational building cases. The Red Centre and the Tyree Building are situated on the UNSW campus in Kensington, Sydney (See **Figure 2**). Both of these buildings allow data to be collated regarding safety, security, and defence at larger scale due to the size of each building. The buildings themselves are open to the public during the day (no swipe access at main entrances and to public parts at each level), and the boundaries are large glass facades on the ground floor. What protects these glass facades from a potential car/truck is a set of stairs between the University Mall and the interiors. The survey enables to make distinctions between both buildings.

As the appreciation of perception of safety on a scale from 1 to 5 is different for each person, we asked the students to compare one location with another. The information about the differences of perception of safety from one location to another is more accurate than the absolute value itself for one location.

The comments added by the students in the semi-structured sections of the survey enable to appreciate the level of anxiety or well-being of the student better: two students defining the level of safety 3/5 might not have the same level of anxiety. The information in the textbox aims to highlight these differences.

To get a better understanding of the level of safety ranked by the students for the entries of the two buildings, we visualised the average perceived safety level for each entry in ArcGIS and for a better communication with users, and to enhance accessibility to the data, we visualised the GIS layer in Google Earth. To map these ranks, we considered a range of 5 to 1 for mapping feeling of safety where 5 demonstrates feeling of distressed using the entry and 1 reveals feeling totally safe using the location as chosen entry.

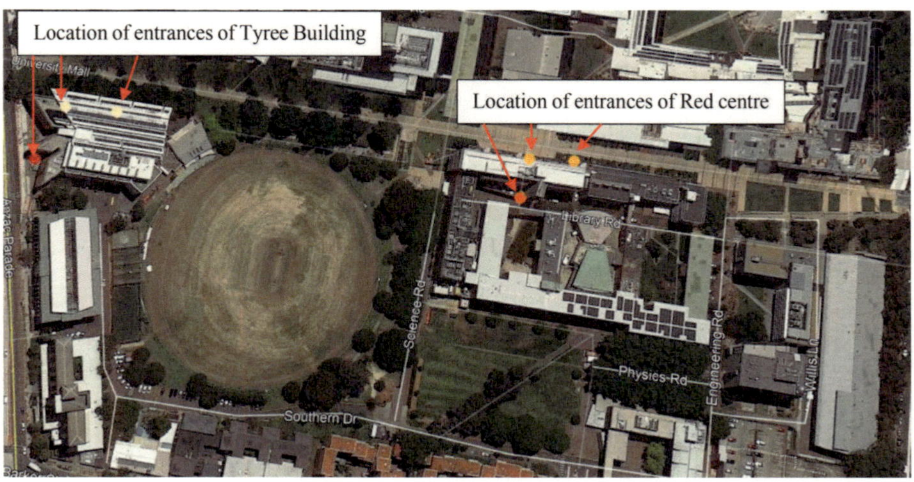

Figure 2.
Red centre and Tyree building at UNSW and the studied entrances in this research.

4. Findings and discussion

Figures 3 and **4** show the feeling of unsafety, for Red Centre and Tyree Buildings entries in different day and night times.

Figures 3 and **4** show the average for each entry gate at different time. **Figure 4** shows that students perceptions in terms of safety for entry 3 (Tyree) are always the same, while entry 1 and 2 may not be safe sometime. Similarly, the same feeling at night time where some lighting facilities are available in front of the case building.

Figure 5 also shows that case 1 has received consistent scores which are average and high. However, the second case entries do not allow students to have a consistent safe feeling at different times, and they may need to change their choices of paths and entries. Based on the evaluation and the available data, the risk map was created indicating the main spots. Average level of safety is demonstrated in different colours in a map created in ArcGIS. The layer is exported into Google Earth, and as can be seen in **Figure 6**, the values of average safety for each entry can be seen and compared. Indeed, this kind of mapping can be used for identification of the problematic areas. In addition, it could be used as a safety base map for the security teams to enhance intervention activities or monitoring facilities for the entries identified to have lower levels of safety.

The facade design was a consistent comment across a number of the students also, suggesting that glass or transparent facades lead to more secure buildings/increased feelings of security. The entrances/exits and circulation were also mentioned in responses from students, suggesting that the design of these has a major impact on the safety of these buildings (particularly in the RC which has a single main entrance/exit point).

The Tyree Building has greater variation across the locations, with Location 03 (from back of building) generally regarded as the least safe option, particularly at night, and Location 02 (from University Mall) regarded as the safest at both times of day. Location 03 is regarded as least safe.

In terms of design we investigated that in Question 9f: Do you think the building layout and architecture (accesses, evaluation of facades, and close surroundings up to 200 meters) have an impact on the level of criminal acts and their nature? The interviewers state that: *The configuration of spaces manipulates places for the human condition and consciousness to use it for their benefits. Criminals acts are not the only factor.*" And also "*the RC has a clear facade and so is quite open. There is also essentially 1 entrance as well.*" Or another participant clarifies that "*the facade of the Red Centre building for the exhibition space and mezzanine are very visible so this may discourage criminal activity; however, the upper floors consist of long corridors and closed off areas may allow for more criminal activity. The concentration of major circulation routes and inefficiency of emergency exits may also encourage this type of activity.*" And finally, "*more advanced designs on facades makes building appear more safe and secure.*"

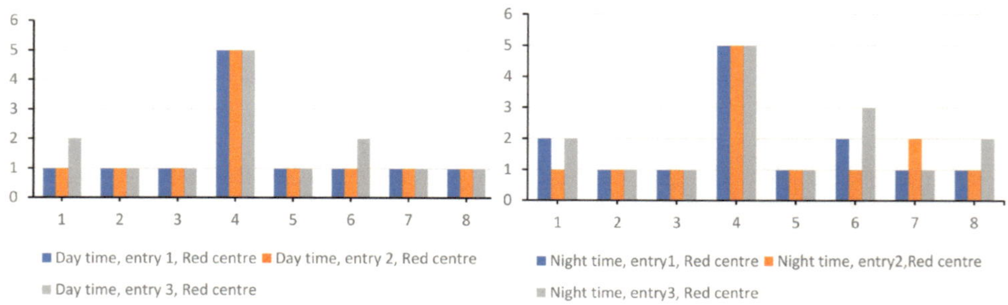

Figure 3.
Day and night time feeling of unsafety for red centre building.

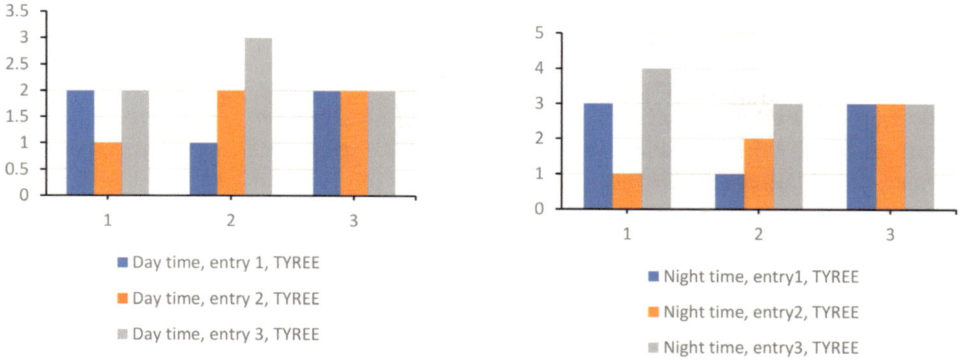

Figure 4.
Day and night time feeling of unsafety for entrances of TYREE building.

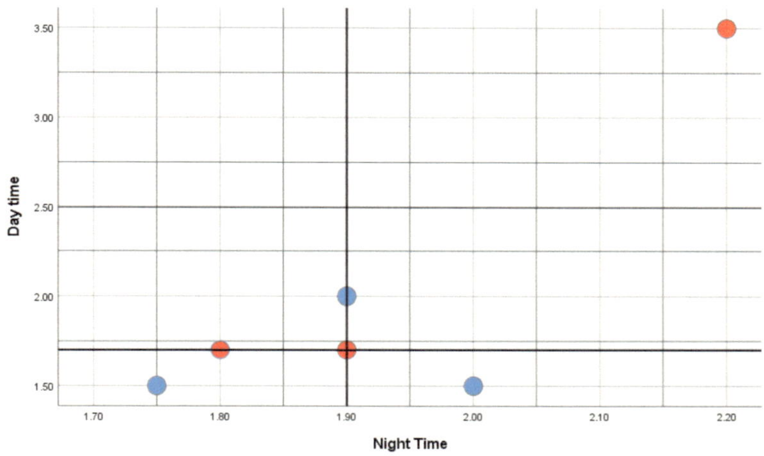

Figure 5.
The average score for each entry gate at night and daytime. Note to figure: Red refers to RC (case building 1) and Blue refers to TB (case building 2). The RC received scores above the median. The TB has received totally different scores for each entry.

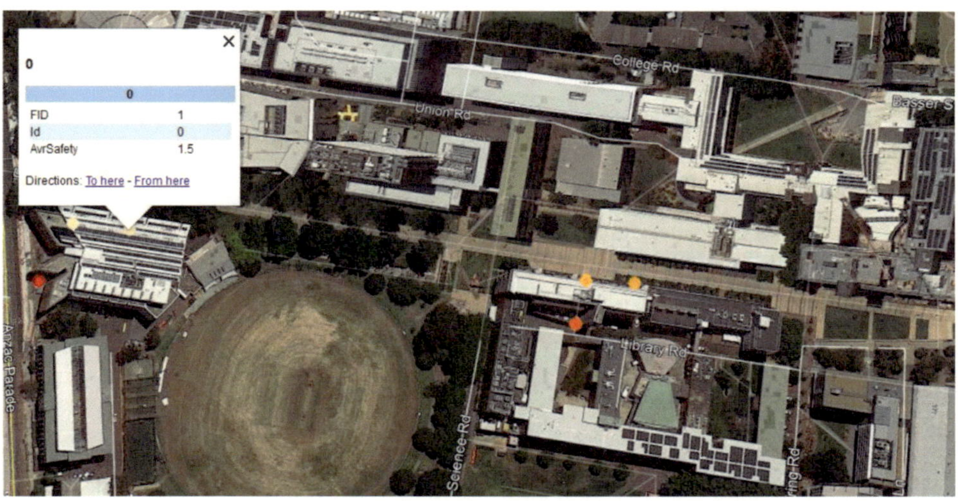

Figure 6.
Demonstration of the rate for a feeling of safety (AvrSafety value in the demonstrated attribute table on the map) for entry 1 of TYREE building.

5. Conclusion and further research

This study aimed to develop a novel method for evaluating and demonstrating risk spots in an educational campus using geographical information systems. The study addressed the following key questions: Which student feels less safe than others, which entry is reported unsafe, and how is daytime safety perceived differently than night time safety? The results show that the Red Centre building is generally regarded as safer than the Tyree Building, during both daytime and night. There are no major differences across both sites during the day; however, the major differences occur for the night rating. Findings indicate that students feel less safe working in the building than the RC work zones. It might be due to the proximity of the TB to the main road (Anzac Parade) and main walkways. One side of the TB is in line with the main road.

Across the two locations, the perception of safety in the RC is consistent during both day and night; however, it was regarded to be slightly safer inside than external location. Across responses to the RC, students either responded with 1 or 5 (extremes of safety or danger), with no responses landing in the middle; it suggests extreme feelings of comfortability on campus or the opposite. Location 01 regarded as most safe, and Location 03 regarded as least safe.

By comparing the results for the two buildings, it is suggested that the openness of buildings is valued by users to feel safe. Glass facades, when designed with safety glass, have a positive impact on the feeling of safety. Safety of university buildings can be achieved through access control coupled with the maintenance of visual links between interior and exterior of the building, i.e. symbolic openness.

The tools developed for this research enable to assist designers and stakeholders in the planning of university buildings. Safety-oriented design, based on the two main strategies of enclosing and layering, can be developed by stakeholders by considering the existing and future users' feeling of safety. Design solutions coupling feeling of safety and architectural quality can be identified and then developed and reused in future projects.

Author details

Sara Shirowzhan, Laurence Kimmel, Mohammad Mojtahedi*, Samad Sepasgozar and Jack Peacock
Faculty of Built Environment, University of New South Wales, Sydney, Australia

*Address all correspondence to: m.mojtahedi@unsw.edu.ua

References

[1] Atlas RI. Understanding CPTED and situational crime prevention. In: 21st Century Security and CPTED. Boca Raton, Florida, USA: Auerbach Publications; 2008

[2] Isdr U. Global assessment report on disaster risk reduction. In: United Nations International Strategy for Disaster Reduction (UN ISDR), Geneva, Switzerland. ISBN/ISSN: 980852698. 2009. p. 207

[3] Huq S, Kovats S, Reid H, Satterthwaite D. Reducing Risks to Cities from Disasters and Climate Change. London, England: SAGE Publications Sage UK; 2007

[4] Reduction ISFD. Living with Risk: A Global Review of Disaster Reduction Initiatives. United Nations Publications; 2004

[5] Kelman I, Gaillard JC, Mercer J. Climate change's role in disaster risk reduction's future: Beyond vulnerability and resilience. International Journal of Disaster Risk Science. 2015;**6**(1):21-27

[6] Australia-New Zealand Counter-Terrorism Committee (ANZCTC). Guidelines for Crowded Places. ANZCTC; 2017a

[7] Homeland Security. FEMA 426 Reference Manual to Mitigate Potential Terrorist Attacks against Buildings. Washington DC: US Department of Homeland Security; 2011

[8] Homeland Security. FEMA 430 Site and Urban Design for Security. Washington DC: US Department of Homeland Security; 2007

[9] HM Government. Crowded Places: The Planning System and Counter-Terrorism. London: HM Government; 2012

[10] HM Government. Protecting Crowded Places: Design and Technical Issues. London: HM Government; 2014

[11] RIBA. RIBA Guidance on Designing for Counter Terrorism. London: RIBA; 2010

[12] Nadel BA. Building Security: Handbook for Architectural Planning and Design. New York City, United States: McGraw-Hill Professional; 2002

[13] Paradis R, Tran B. Whole building design guide: Balancing security/safety and sustainability. USA: National Institute of Building Sciences; 2007

[14] Division of School Support North Carolina. Design of Schools to Prevent Violent Attack. Raleigh, NC: Division of School Support; 2008

[15] Australia-New Zealand Counter-Terrorism Committee (ANZCTC). Crowded Places Self-Assessment Tool. ANZCTC; 2017b

[16] Demkin J. Security Planning and Design: A Guide for Architects and Building Design Professionals. Republished. Michigan, United States: John Wiley and Sons; 2003

[17] Ferreira J, João P, Martins J. GIS for crime analysis-geography for predictive models. The Electronic Journal Information Systems Evaluation. Portugal; 2012:**15**(1)

[18] Santos RB. Crime Analysis with Crime Mapping. California, USA: Sage Publications; 2016

[19] Shafique I, Zahra SA, Farid T, Sharif M. Role of GIS in crime mapping and analysis. Sukkur IBA Journal of Computing and Mathematical Sciences. Pakistan. 2017;**1**:39-47

[20] Sheikh J, Shafique I, Sharif M, Zahra SA, Farid T. IST: Role of GIS

in crime mapping and analysis. Communication technologies (Com Tech). In: International Conference on, 2017. IEEE; 2017. pp. 126-131

[21] Chainey S, Tompson L. Crime Mapping Case Studies: Practice and Research. New Jersey, USA: John Wiley and Sons; 2008

[22] Wilson R, Smith K. What is applied geography for the study of crime and public safety. Geography and Public Safety. 2008;**1**:1-3

[23] Johnson C. Crime Mapping and Analysis Using GIS. Geomatics Conference on Geomatics in Electronic Governance. Tallin, Estonia; January; 2000

[24] Shirowzhan S, Lim S, Trinder J. Enhanced autocorrelation-based algorithms for filtering airborne Lidar data over urban areas. Journal of Surveying Engineering. 2016;**142**:04015008

[25] Shirowzhan S, Sepasgozar SME, Zaini I, Wang C. An Integrated GIS and Wi-Fi Based Locating System for Improving Construction Labor Communications, International Symposium on Automation and Robotics in Construction; 2017. pp. 1052-1059

[26] Neal A, Griffin MA. A study of the lagged relationships among safety climate, safety motivation, safety behavior, and accidents at the individual and group levels. Journal of Applied Psychology. 2006;**91**:946

[27] Bosworth K, Ford L, Hernandaz D. School climate factors contributing to student and faculty perceptions of safety in select Arizona schools. Journal of School Health. USA. 2011;**81**:194-201

[28] Sundstrom E, Herbert RK, Brown DW. Privacy and communication in an open-plan office: A case study. Environment and Behavior. 1982;**14**:379-392

[29] Dowling R, Atkinson R, Mcguirk P. Privatism, privatisation and social distinction in master-planned residential estates. Australia: Urban Policy and Research. 2010;**28**:391-410

[30] Loukaitou-Sideris A. Safe on the move: The importance of the built environment. In: The Urban Fabric of Crime and Fear. Springer; 2011

[31] Räty TD. Survey on contemporary remote surveillance systems for public safety. IEEE Transactions on Systems, Man, and Cybernetics, Part C (Applications and Reviews). 2010;**40**:493-515

[32] Wang X. Intelligent multi-camera video surveillance: A review. Pattern Recognition Letters. 2013;**34**:3-19

[33] Rothrock SE. Antiterrorism Design and Public Safety: Reconciling CPTED with the Post-9/11 City. Massachusetts Institute of Technology; 2010

[34] Basile O, Persia L, Usami DS. A methodology to assess pedestrian crossing safety. European Transport Research Review. 2010;**2**:129-137

[35] Lee JM. Automated Checking of Building Requirements on Circulation over a Range of Design Phases. Georgia Institute of Technology; 2010

[36] Kobes M, Helsloot I, De Vries B, Post JG. Building safety and human behaviour in fire: A literature review. Fire Safety Journal. 2010;**45**:1-11

[37] SS-SIS-RP A. Equipment and Technology for Public Transit. USA: American Public Transportation Association; 2013

[38] Richardson C. Working Alone: Protecting and Building Solidarity in the Workplace of the Future. Boston, USA: Labor Resource Center Publications; 2010

[39] Keval H, Sasse MA. "Not the usual suspects": A study of factors reducing the effectiveness of CCTV. Security Journal. 2010;**23**:134-154

[40] Jones MK. Interactive Personal Surveillance and Security (IPSS) System. Google Patents; 2013

Chapter 5

The Effect of Place Attachment on Educational Efficiency in Elementary Schools

Farhad Soheili, Reyhaneh Karimi, Behnaz Avazpour and Samad M.E. Sepasgozar

Abstract

Close attention to education and its effects on the identities of societies has been ingrained in the history of studies. Nowadays, a considerable amount of criticism in the field of education is related to inefficient and obsolete educational scheduling, with a large number of researchers exploring new learning methods. Logically, these new methods require new architectural environments, which can meet the needs of a new generation with the approach of positively affecting students' learning efficiency. The main purpose of this research is to identify effective factors in students' sense of attachment to school. First, this research finds factors on the sense of attachment by studying past literature. Second, a number of questions and hypothesis are developed. Third, in order to validate the effective factors and answer the developed questions, a questionnaire is devised and distributed among 278 students of 10 state primary schools of Shahriyar, Iran. The collected data are then analyzed using SPSS software. The findings indicate that there is a meaningful relation between students' educational efficiency and their sense of attachment to school. It also shows that important factors, such as social activities and connection with natural elements in design, can have a noticeable impact on the students' sense of attachment to school and consequently their educational efficiency.

Keywords: elementary school, place attachment, sense of attachment, educational efficiency

1. Introduction

In recent years, attention has been drawn to the importance of education in developing countries. The reason for this is a rise in global awareness of this issue. Each year, many countries in the Persian Gulf region allocate significant expenditure toward improving schools and universities according to the most up-to-date methods and equipment. In Iran, the conditions seem to be different. In spite of the formation of various educational organizations, each with a systematic program to improve the quality of education, the situation in reality seems to be different. In order to clarify the subject, we propose a few questions:

- Do Iranian students go to school every day with the interest in new lessons and teachings?

- Do the environments of schools lead students to become successful people?
- Are the thought concepts in schools permanent and effective, or are they forgotten after completing the exam period and do they remain intact at the community level?
- What are the key issues in students' effective learning? Are they high self-esteem, high sense of collective participation, creativity, etc.?

Answers to the above questions are critical in better understanding the learning process and its impact on individual student identities. According to many scholars, the identity at early ages is the foundation for the future identity of communities.

Students attending school early in their childhood typically show enthusiasm for learning; however, after a short period of time, the passion suddenly subsides. They spend time in school environments that do little to increase their willingness to learn. It is likely that such environments have been stagnant in their pattern of construction for almost half a century and have not, in essence, been successful even with the development of its educational programs. The importance of addressing this issue from a variety of perspectives must be considered. At the current time that many institutes in different parts of the country are building schools in high numbers, the quality of these schools and the impact they make on the future of each individual and, more generally, on society require close attention.

Students and children are a very important factor in building and making progress for any society and the way they make decisions can impact the society, which is vital by various institutions in today's developed countries. And to attain the ability to compete with other communities, children are instructed and trained in schools that implement appropriate and useful approaches in educational program areas as well as in school architecture design. It is obvious that the designed environment has the potential to affect individual behaviors and can continue to influence personality through the continuation of the same process. Therefore, it is important to address the issue of building schools by adopting a suitable approach focused on designed environment.

By their nature, individuals' identities form and vary throughout their life. Children between the ages of 6 and 9 are at a critical period in this personality formation and find their character comparable to others [**1**]. Hence, it is expected that the role of elementary schools in children's lives during these years will be given greater emphasis and can have a positive impact on the formation of their individual identities.

The main purpose of the present research is to extract effective factors in creating a sense of attachment in students to the school. Such factors, instrumental in designing and creating a sense of attachment, can have a positive impact on students' educational efficiency. By asking questions such as "Does increasing the sense of belonging to the school increase educational efficiency?" and "What factors in school design can increase the sense of attachment to the school?", the study explores the following hypothesis: "the sense of attachment to school has a positive impact on their educational efficiency."

2. Literature review

In reviewing the past studies on designing educational spaces, and in particular the design of schools, it is evident that designing schools has been of great importance to the researchers for many years. John Dewey addresses the subject about a century ago, stating that the environment of schools must be considered as one of the primary societies. He further emphasizes the importance of these types of societies. In his view, the general atmosphere governing the school environment and the activities that take place can directly reflect the general public's perceptions [2].

Burke and Grosvenor in a book entitled "The School I Love" were trying to find the school-specific characteristics of their students. The first line of this book begins with questions that ask about the ability of schools to meet the needs of children and adolescents. In this book, which is full of children's views in the form of poetry, painting, and interviews, there are some points about the interaction between children and school environment, their expectations of learning, and the problems of educational systems [3]. The Guardian magazine hosted a contest entitled "The School I'd Like" in 2001 and invited children to comment on their favorite school. The school we like is this: a beautiful school, a comfortable school, a safe school, a listener school, a flexible school, a school associated with us, a respectable school, a school without a wall, or a school for everyone. Burke and Grosvenor discuss school buildings, classes, courtyards, and other spaces, and comments from students are presented in each section, after reviewing relevant studies by scholars. Many of these comments refer to the extracted subsets of the sense of belonging to the school space. For example, an eight-year-old girl named Hannah describes her school as: "The school I love is a beautiful park in the valley passing by a river" [3].

This description notes the interest of students in communicating with the natural environment, called the "interaction of the active environment."

School buildings, their environments and landscapes, open spaces and classrooms, layout, and decorations all are referred to as the third teacher [4]. In the opinion of many scholars, there is a need for school to be a guide, and a part of it should convey some of the shoulds and shouldn'ts to the students [5].

In this context, the term "school exemplary" has been used by a number of scholars for leading schools. In a research conducted by Christopher [6] with the aim of extracting the characteristics of primary schools, ten factors are introduced as the main characteristics of these schools. One of these factors was the performance of the school building as a teacher, using design elements in structural units and facilities, which would have made students curious about how they work. Another feature of these schools was to provide an environment where users would feel at home, as well as to develop a sense of participation in social activities in their students, thereby strengthening the sense of being one and uniting them [7, 8]. The characteristics of these schools are listed below:

- The buildings of these schools are designed according to the environment and are part of its assets.
- Paying attention to details of building construction up to its smallest components.
- Various space experiences for social and individual activities.
- Designing spaces that enhance social skills.
- Providing enough space to provide services to all students.
- Flexible design.

In a study published by the Thomas Jefferson Foundation, it was found that school design can push students to positive behaviors on their own, and as students derive this spirit from the school space, opportunities for positive influence in their education and learning are created [5]. He also says that children are highly capable of observing, and any shortcomings that designers may have in design can give them the feeling that they are not essentially in an important place.

The comments mentioned above have long concerned researchers trying to improve the educational environment and have tried to distinguish the building of schools from barracks and give them an appropriate identity, and transfer this distinction to the students, an identity that can in turn create positive effects on student behaviors [9].

In a study by Maxwell and Chmielewski, the role of classroom personalization by students and its impact on students' self-esteem were explored. The results of their research indicate the positive effects of classroom personalization on self-esteem and personal identity of primary school students as well as children in kindergarten. In their findings, the authors note that the presence of preset spaces in order to enable students to visualize their works can affect their self-esteem [1].

Ma, in a research on the sense of attachment to the school, states that there is a direct relationship between students' self-esteem and their sense of attachment to the school. In fact, their view of themselves can be generalized to their view of the school environment. In this regard, the school environment and its design can be affected, and designers should be diligent in designing the school in order to increase students' knowledge of their individual identities [10].

In many studies in the context of the psychology of the environment in relation to educational spaces, researchers have sought to identify the factors that enhance the sense of self-esteem in students. In a study entitled "The Role of the Continuing Presence of Students' Effects on Their Sense of Self-Esteem," it was found that students who were involved in class layout and decoration, and had their works displayed continuously, had a higher sense of self-esteem and have shown a more appropriate educational performance [11].

By studying four schools, Upitis, examines issues that increase the sense of belonging to the school. In his view, the school environment should be seen from two aspects of the natural environment and artificial environment, an example of which is shown in **Figure 1**: a part that is the main building of the school and a part that is the natural environment where students attend school. By examining case examples, he shows that the presence of nature and the provision of direct student interaction with it can have a very positive impact on increasing their sense of attachment to the school environment [12].

Figure 1.
Artificial and natural aspects of a school environment located in Shahriyar, the photo is taken by authors.

The school should be a place where students acquire their personal identity, where educational process takes on a meaningful form, where the educational experiences students receive turn attractive to them, and where the learning environment is welcoming to all. Collectively, such attributes give a particular morale to the school's atmosphere, which is subsequently passed onto students [5].

In many writings of scholars on the sense of belonging to the school, the role of teachers has been emphasized in the creation of this feeling. At this stage, it is important to note that the appropriate school environment is not just for students and it affects the behavior of teachers and their training process [13]. This positive impact can, in effect, directly affect students' learning and also create a sense of belonging in them.

In the area of school design, a significant part of the research focused on the impact of the quality of the educational environment on the student's educational situation. The researchers also conducted independent classroom research and attempted to illustrate the interaction among the classroom conditions, educational condition, and students' behavioral conditions. Their studies indicate that the class environment, depending on its type of design, can have a positive or negative effect. On the one hand, the classroom environment can lead to positive educational and behavioral improvements; on the other hand, the classroom may also increase inappropriate behaviors, increase the desire for isolation and avoiding social interactions in school, as well as increase stress among elementary students or preschoolers [14–16].

Knight and Noyes conducted studies on classroom space in relation to the relationship between class arrangement and student behavior. In their view, such a study was necessary because most students spend time inside the classroom; that is, the space, along with its furniture, can play a significant role in students' behavior. A sample of class design with various furniture types is shown in **Figure 2**. Comparing the old chairs and new chairs that entered a school, the authors found that chairs were essentially requiring students to do an activity and its newness per se cannot be efficient and useful [17]. Other studies regarding classroom arrangement-behavior include the study of the effect of noise pollution and density in the classroom and their role in student learning [18–21].

Another part of the study is the school's overall atmosphere. In a study entitled "Explaining the Relationship between Achievements of the Students and the School's Physical Environment," the basic assumption was that the pattern of school environment design influences students' achievements. In this research, five design patterns including access, open spaces, visibility and perspective, access to natural light, and finally the way of placing classes together, were considered as design

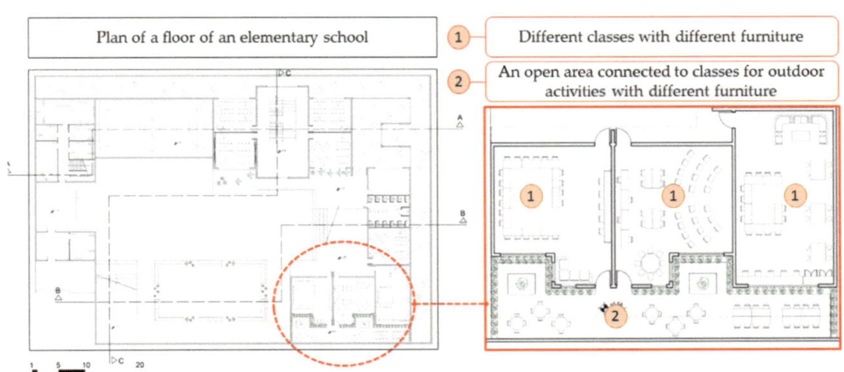

Figure 2.
Various furniture types designed for classes, by authors.

features. The ITBS test, designed and evaluated to assess students' learning at school, was used to measure students' progress. In this test, the various skill levels of the students can be evaluated. One of the main advantages of this method is the possibility of comparing students' educational situations with their peers and, consequently, comparing one school with another school. The results of this study revealed a very significant relationship between the type of school design and the student's learning. This study puts an emphasis on further research on the field of learning, since there are another key factors contributing to the mentioned field that need to be investigated, such as: the status and ability of school teachers and their skills in teaching, students' economic and social conditions, racial and ethnic differences, and many other factors [22].

Over the last few decades, the issues of educational spaces and school facilities have been raised in scholarly debate. Specifically, there was some discussion about the necessity and importance of building new schools. The origin of these concerns, on the one hand, was economic issues and, on the other hand, with increasing population growth, was sufficient time to build. The main question was: could old schools be restored to new schools instead of building new schools and adding new ones? McGuffey studied seven case studies and, in almost all cases, students whose schools had a lower age of construction and had newer equipment showed better performance [23].

Philips noticed a significant difference in reading and math skills among students in new buildings and those in older schools. His research revealed a clear relationship between the age of the building and its equipment with the quality of students' performance. Students who were in the newer building and environment, with more up-to-date facilities, showed a 7.63% increase in their grades [24].

Another issue that has been addressed by researchers is the impact of school size and capacity, as well as the size of the classroom on the student's educational situation [25]. In his 2008 study in the Danish context, Duke divided schools into two large and small groups. At high school, both groups of schools had their own positive benefits [26]. Having said that, Cotton found that in small schools, student achievement is sometimes more dramatic than that of older school students [27]. Of course, both scholars in their writings state that the smallness and largeness scale of schools vary from one region to another.

In the elementary school, this has been given less attention, because the general belief is that smaller schools provide a more suitable atmosphere for the elementary school. However, in a study of seven schools in California, schools with more students than those set for elementary schools were better educated than the smaller ones [28]. However, it seems that this requires more in-depth research.

Another topic that has been taken into consideration by researchers is outdoor education. Over the past few years, this has attracted more attention and led to the design of schools without walls. Research shows that students regularly recall times that they spent outside the classroom framework as academic visits or student camps. Falk and Deirking admit that 96% of the subjects in the study group referred to the memories of their scientific journeys in their early years. He goes on to say that it's not just a matter of remembering these times to show that they are useful, and that even its usefulness does not result in the inappropriateness of training in ordinary classes [29].

Other researchers have found that educational experiences transmitted to students in open environments are far more influential than classroom environments [30]. This effect is essentially based on the desire of students to learn from the outdoors. But this interest varies over the age range. Research shows a significant difference in the tendency of primary and secondary school students to learn in open spaces. Elementary school students were significantly more enthusiastic than high school students after experiencing a scientific journey [31]. Outdoor education can provide activities that are not possible in the classroom in the form of an outdoor training workshop. Exterior spaces are environments suitable for activities that deal with natural elements such as water, soil, and mud and can increase student motivation; a sample of such a place is shown in **Figure 2**.

As observed and reported, many researchers have investigated various sectors of designing a school. Such studies, each with a different approach, examined a part of the subject of education and the role of architecture in student learning. In a number of these studies, there are common points in the research approach and the subject matter.

In the area of Iranian research, scholars have also focused on designing schools based on different approaches:

Some researchers mentioned several interesting points in an article entitled "Recognizing the Role of Nature in Educational Spaces." They emphasize the role of nature as a factor in the attractiveness of school education to different aspects of such an impact. From their point of view, examining different aspects of this impact could lead to more diverse designs. They eventually suggest design patterns such as creating open spaces and interacting with natural elements, creating eye-catching dimensions at different points, using open spaces in class design, as well as student changes in designed gardens [32].

Kamel Nia explores the factors for better environmental quality in schools in a book entitled "Learning Environment Design Grammar." He states that because most educational environments form a narrow line of view for students, it is necessary to use the horizontal elements to create the appropriate visions and perspectives along with establishing a more effective external and internal interaction [33]. **Figure 3** represents an example of the mentioned elements.

In this regard, another researcher in an article entitled "The concept and function of open space in traditional and new schools," historically addresses the different roles of these spaces and uses the model of central life as a successful and effective model of outdoor education [34].

A study entitled "Space perception; a key element in designing children's spaces" presents the views of children gathered in the form of painting and opinion polls. The results show that children essentially paint their school in interaction with nature, and the details of the chairs and benches, the blackboard, and the overall form of building the school are very important to them. Also, in terms of the colors used for the school building, the children are more pleased with the lighter and brighter colors for their classes [35].

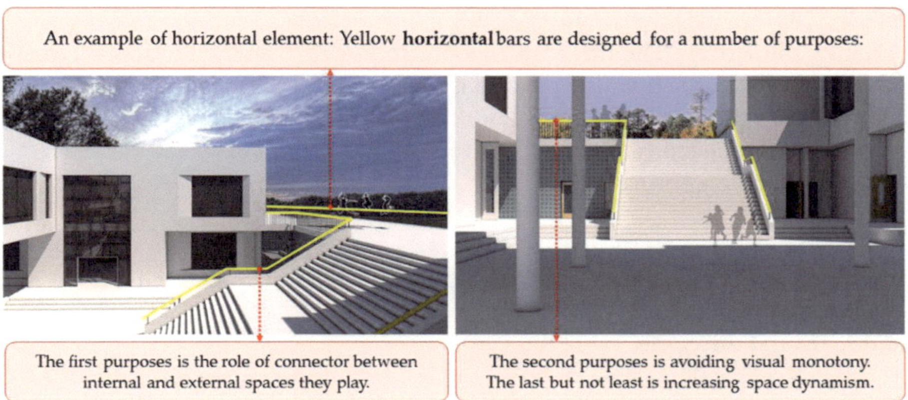

Figure 3.
A sample of horizontal element, designed by authors.

The issue of flexible design in educational spaces is explored in an important study entitled "Flexible learning environment." In the researchers' view, the advantage of this type of design is the ability to respond to the different needs of children in their childhood. On the other hand, it is possible to adapt to the ongoing changes in educational programs that require different spaces. Subsequently, with the introduction of the ideal class, they set forth factors of spatial fluidity, internal and external communication, communication with the community, spatial transparency, and livelihood of communication corridors and transitional spaces to make the design more flexible [36].

"Criteria for Designing Future Education Space" investigates the changes needed in design attitudes in the new age. In this article, it is pointed out that the architecture of today's schools is very similar to its original examples in the country and has not adapted itself to changing lifestyle and educational system. Then, he identified and studied the factors needed in the field of school education and architecture, and tried to provide solutions for architectural harmony and implementation of new educational ideas in schools. One of the characteristics that is mentioned at the end of this article is the issue of flexibility in design, which can be adapted to the new educational needs of the curriculum as an important factor [37].

The article "Study of the environmental components of learning space and its impact on student behavior" states that humans are always interacting with their environment and this interaction is two-way. That is, people affect the environment and the environment affects people. The environment is effective due to its structure in shaping the personality and physical and mental development of each individual. One of the environments in which human beings spend most of their lives is a learning environment. The physical space of schools also affects the behavior of children, which is the future capital of our country, and, as a result, a lack of attention to design can create the background for abusive behaviors in students. This research seeks to answer the fundamental question of how the physical environment of educational environments affects the behavior of children. In this regard, understanding the concept of the environment and the interaction between the child and the environment, we first point out the physical components of the educational environment and their impact on the cognitive processes and social behaviors of the students [38].

The article entitled "A Study of the Way to Create a Sense of Place for Restoration of Identity and Confidentiality in Architectural Areas" focuses on space and location as a theoretical base in various fields. Its fundamental findings include achieving the recognition of space and methods for semantic quality and its transformation into a place and, consequently, the creation of a sense of place and place belonging [39].

Finally, the article "The Effect of School Architecture on the Students' Learning Process" states that the learning process is ongoing, and this occurs naturally in a space, and since the environment influences human beings, it can affect the learning process too. The study highlights factors such as light, color, and furniture affecting the learning process [40].

3. Theoretical framework of research and key words definition

3.1 Educational outcome

Training is a purposeful and predesigned activity aimed at providing opportunities to facilitate and speed learning in a breeding system. Training may be conducted through films, radio, television, and other media, with or without the presence of a professor [41]. The meaning of the word efficiency in the dictionary comes as a result of work and efficiency. Therefore, the goal is to increase the educational outcome and improve the outcome of the work and the learning activity that takes place in learning.

3.2 Elementary school

The school is an educational organization that is usually located in a building, based on official standards and certain criteria. By education system supervision, the school is responsible for implementing approved programs, offering a variety of education and training for students at the level or course of study. After graduation, they will be awarded an official degree. The elementary school is a six-year course that students enter at the age of seven. After completing the first round (first, second, and third grades) and the second round (fourth, fifth, and sixth grades) and if they succeed in the exam, they will receive a certificate of completion of the elementary school. Public schools are also schools where students can study for free, and students can study at these schools if they have the required age and residence range.

3.3 The sense of place attachment

The sense of place attachment is defined as the emotional connection between the individual and the place [42], which Shumaker and Taylor refer to as a positive or influential link between individuals and their environment [43]. Hummon introduces it as an emotional involvement with the environment [44]. Rubinstein argues that the positive experience that a person acquires from the place is the result of the positive emotions that are formed between the individual and the place. The more emotional the relationship is, the more interested a person is in the place [45]. One of the things that is constantly observed in all studies about the sense of place attachment is the presence of emotion and excitement at the heart of these definitions. The next feature of these definitions is "the desire to stay close to the subject to which it belongs." This attribute is implicit in many of the definitions available. By combining this feature with the definitions mentioned above, it can ultimately be reached: an emotional and positive relationship between a person and a particular place that leads to motivation and staying in that particular place [46, 47].

3.4 Factors affecting the sense of place attachment

3.4.1 Cognitive-perceptual factors

Indeed, the greater the individual's awareness of the place, the more the sense of place and attachment to the place can increase. An important factor in this case is time, which means that over time, the person becomes more familiar with the environment and gets a better understanding of the place. The other two factors are related to the following perceptual factors:

3.4.1.1 Spatial and physical distinction

School design with a variety of spatial qualities, as well as the use of materials of varying color and texture, can add to the identity of each part of the overall space. This affects the students' perception of the space that they are involved in and increases their sense of belonging to the place [48].

3.4.1.2 Environmental legibility

Another factor that positively affects the person's perception of the environment is the increase in the level of individual attachment to the location, which is the cause of the readability of the environment. In his book [49] on the sense of place attachment as one of the most effective features in design, Kevin Lynch poses the issue of environment legibility and examines the factors that make this legibility. Lynch goes on to outline features and body elements, paths, edges, cues, and nodes as effective in shaping the concept of legibility of the environment [50].

3.4.2 Social factors

As previously stated, a sense of belonging can be the result of the individual-place interaction or the individual-individual interaction in a particular place. In fact, just as cognitive factors play a role in creating a sense of place attachment, social factors can also contribute to the formation of this sense. As mentioned above, past research has shown that the social dimension of the sense of attachment is stronger than its physical dimension. Therefore, the environment that makes these interactions possible is an effective step in creating a sense of attachment. One of the needs that humans have always sought to answer is to have a proper definition of their individual identity. Social interactions, in the context of communicating with family, friends, and relatives, provide an important part of the context for responding to social needs and acceptability [51]. This need for children is also demonstrated through communicating with parents and other family members, teachers, and other adults as well as other children [52].

3.4.3 Environmental-physical factors

3.4.3.1 Personalization of the environment

Existence of personal elements in the environment is another factor in strengthening individual identity, especially that of students. For example, providing spaces, in which students can display their artwork on a permanent basis like a sample that is shown in **Figure 4** or the presence of personal closet and other spaces

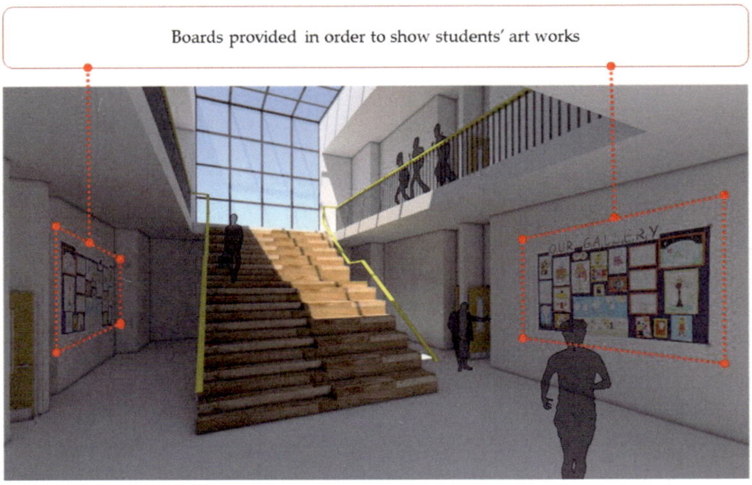

Figure 4.
Permanent galleries, where students can exhibit their art works, designed by authors.

and equipment that convey a sense of ownership of part of space to students, can significantly enhance a student's sense of attachment to school. This is one of the environmental-physical factors that develops a sense of attachment to a school's physical dimensions. Riger and Lavrakas also emphasize the role of these physical factors in their writings [53, 54].

3.4.3.2 Connection with natural environments

One of the most important factors and perhaps the most effective factor in creating a sense of attachment to a site and, specifically, the school site is the use of natural elements in design. Likewise, many researchers point to the importance of this topic in the field of environmental psychology, and it is referred to as a factor in shaping the active environment interaction, which can furthermore lead to a sense of attachment [55]. Also, in designing educational spaces, many studies and patterns direct designers to the use of natural elements in the design and creation of spatial fluidity between the inside and outside and the natural and artificial environment [56].

3.4.3.3 Answering the functional needs

If an environment does not provide functional requirements to the people occupying it, to whatever extent it offers factors of a sense of attachment, it still fails to perform its duties toward the individual. This is particularly evident in the design of places like the school, which remains a central issue for the users.

3.5 Theoretical research framework

According to the main hypothesis mentioned in the research, "the sense of place attachment for students in relation to the school has a positive impact on their educational efficiency," and the sense of attachment to the school space is considered as an independent variable. The rest of this research attempts to prove that it is effective on the dependent variable, which is the efficiency of the training. **Figure 5** shows the relationship between variables and their components.

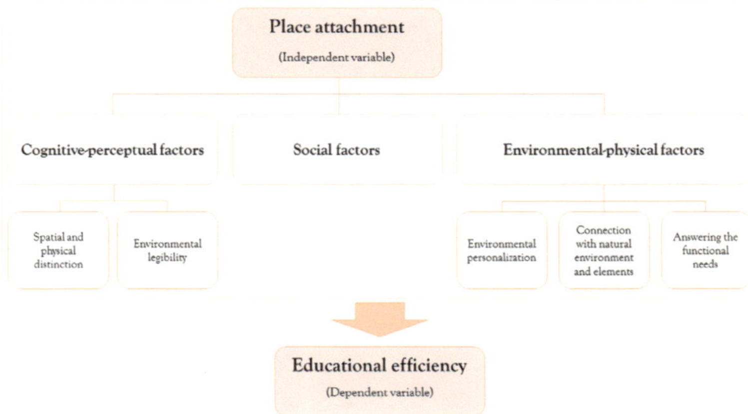

Figure 5.
Theoretical research framework.

4. Research methodology

So far, we have reviewed the extant literature in the field of education, considering the relevant definitions, general assumptions, and the design of questions. In these studies, there is a great deal of difference in the attitudes toward education, school curriculum, and school buildings in developed and developing countries, as well as regarding the measures taken in these areas within the country. Therefore, the need to assess the factors that make sense of attachment to the school is evident. Also, the extent of the impact of the school's environment and its relationship with students' educational efficiency needs to be investigated. In this regard, following the extraction of factors affecting the sense of attachment, based on the extant literature and the goals of the current research, a questionnaire in the form of online survey was created. In line with these aims, questions were put together and distributed among students in a questionnaire. In order to investigate the relationship between students' sense of place attachment and their educational efficiency, the questionnaire was distributed among three groups of students with high, moderate, and poor educational status. Creating this category allows communication between students' sense of attachment and their educational efficiency. Information extracted from the questionnaire and its analysis can lead to the presentation of useful factors and patterns in school design.

The statistical population of this study is elementary schools in Shahriyar city, which consists of 10 schools. All schools selected were public schools, which offer a higher homogeneity of the schools' spatial quality allowed for a broader generalization in the next step and also for extracting more relevant information. Another factor that was used as far as possible in the selection of the statistical society was the economic situation in the families of these students. This was made possible by choosing a certain range of the city of Shahriyar and also considering the fact that these schools are public schools. In order to eliminate the gender factor and its effect on the sense of place attachment, all of these schools were selected from boys' schools. **Figure 6** shows some parts of a selected school.

With regard to similar research and the process that other researchers have taken, it is important to note that in cases where there was a need to measure the degree of attachment in elementary students, generally, the third or fourth to sixth grades were studied. The reason for this was the lack of trust in the views, as well as the lack of space and time that students had in the first, second, and third grades in

Figure 6.
Images of different parts of a school located in Shahriyar, taken by authors.

schools. These younger students generally exhibited the initial excitement of attending school; however, their views and enthusiasm could change reaching higher grades. This, of course, does not mean that they are generally discredited; it is merely a decision that has been made by other researchers in terms of research methodology. Therefore, students of the second elementary school were selected so that more information could be extracted from the questionnaires.

To determine the sample size using the Morgan table and according to the number of students in the statistical society, the population size was 278. The questionnaire was distributed among these students.

The validity and reliability of the questionnaire used in the present study were evaluated using the views of architectural professors as well as primary school teachers in order to achieve the best results.

To assess the reliability of the questionnaire and to examine the correlation between questions that asked common factors, Cronbach's alpha test was used in SPSS software. After applying the Cronbach's alpha test, values above 0.7 were acceptable (**Table 1**). The value of Cronbach's alpha is as follows:

The general hypothesis and questions that shape the purpose of the present research are as follows, respectively:

1. What factors in school design increase students' sense of attachment to school?

2. Does enhancing the sense of attachment to the school increase the educational efficiency?

    - Students' sense of attachment to school has a positive impact on their educational efficiency.

Further detailed questions and hypothesis examining the theoretical foundations associated with the sense of attachment variable, and including a wider section of the main hypothesis and question, are the basis for the preparation of the questionnaire distributed among the statistical population in question. The information extracted from this questionnaire will highlight important factors in school design that will give students a sense of place attachment and increase their educational efficiency. The findings of the questionnaire will also provide a snapshot of the current relationship between the degree of students' interest and their sense of attachment to the school and their educational efficiency. The subhypotheses are as follows:

1. The existence of a personal locker for each student in school increases the sense of attachment to the school and, consequently, increases the educational efficiency.

2. The absence of favorable environmental conditions in the classroom (comfortable temperatures and appropriate light) reduces interest in participating in classroom activities and negatively impacts educational efficiency.

Variable	Cronbach's alpha value
Reliability	0.78

Table 1.
Cronbach's alpha value.

3. Providing cozy spaces and offshoots outside the classroom to study or perform personal activities of students increases their sense of attachment to the school.

4. The presence of green space in the school (in the schoolyard, inside the building, and in classes) can increase the sense of attachment to the school.

5. Open spaces within the school for students' gathering and group activities can increase students' interactions and increase the sense of attachment to the school.

6. Classes that have larger windows with a suitable perspective are more appealing to students.

7. The lack of clarity of class design, the closure of its space, and the lack of visual communication with the outside create unpleasant feelings among students.

The subquestions are as follows:

1. Does the personalization of the environment and having a permanent closet for students throughout the year increase their sense of attachment to the school?

2. Does transparency in the design of school buildings and the interconnection between closed and open spaces increase the perception of space and the sense of attachment to the school?

3. Does creating a spatial distinction and providing spaces for small and individual groups in the school affect student interest in school?

4. Are public spaces more influential in increasing student interactions and enhancing their sense of attachment to the school?

5. Does the presence of green space and vegetation in the school environment (courtyards, public spaces, and classes) increase their sense of attachment to the school?

6. Does classroom environment inappropriateness (sufficient light and comfort temperature) have a negative impact on students' educational efficiency?

4.1 Data analysis

In order to analyze the information obtained in each section, according to the hypothesis formed, a suitable method is chosen for each one and after analyzing it using a selective approach to accept or reject the hypothesis, the analysis is performed.

For displaying descriptive data, numerous tables and percentages are used, and for inferential data, analytical tests are used as follows:

The first subhypothesis was analyzed using one-dimensional chi-square test, the ninth hypothesis was analyzed using the Spearman test, and the rest of the hypotheses were analyzed using the single-variable chi-square method. To exclude more accurate analysis, each section of the analysis of hypothesis includes intersection tables as well as associated diagrams. All data import operations were performed and tested in SPSS software.

A summary of the analysis methods carried out is shown in **Table 2** as follows:

Type of statistical method	Application	Statistical methods and tests
Descriptive statistics	Describing demographic variables, describing main questions and variables	Frequency and rate of frequency
Inferential statistics	Testing hypothesis	Spearman correlation, chi-square test, single variable chi-square

Table 2.
Summary of statistical methods.

4.2 Descriptive statistics

4.2.1 Qualification level of students

To create the possibility of analyzing the relationship between students' educational efficiency and their sense of attachment to the school, three groups of students with a high, moderate, and poor level of education were selected; the frequency of which is shown in **Table 3** as follows:

Students' educational level	Number of students
High educational level	93
Moderate educational level	90
Weak educational level	93

Table 3.
The frequency of students' educational level.

As can be seen, the number of students at different levels is roughly equal and the possibility of comparison between these three categories is provided.

4.2.2 Grade

The questionnaire was distributed among boy students of elementary school (in the fourth, fifth, and sixth grades). The frequency of students in different grades is shown in **Table 4** as follows:

Students' grade	Number of students
4th grade	83
5th grade	103
6th grade	90

Table 4.
Frequency of students in different grades.

4.2.3 Inferential statistics

In this section, inferential findings will be considered. These findings will lead to analyzing hypothesis by using the statistical methods in the SPSS program. Next, the hypothesis is expressed and, using tables and charts, each of them is analyzed.

4.2.4 Theories

1. Analysis of the first hypothesis: "The existence of a personal closet for each student in school increases the sense of attachment to the school and, consequently, increases the educational efficiency"

In this assumption, we examine the relationship between personal closets for each student and their impact on their sense of attachment.

The test used in this case is a single-valuable Chi-square test; the results of which are shown in **Table 5** as follows:

Students' educational level		**Personalization (personal closet)**			**Total**
		There is no difference between having a personal closet in school or not.	**If there is a personal closet in school, it will not be bad.**	**Yes, I like to have a personal closet in school.**	
High	Count	47	13	33	93
	Expected count	28.3	17.2	47.5	93.0
Moderate	Count	17	19	54	90
	Expected count	27.4	16.6	46.0	90.0
Weak	Count	16	21	56	93
	Expected count	28.3	17.2	47.5	93.0
Total	Count	80	53	143	276
	Expected count	84.0	51.0	141.0	276.0

Table 5.
*Intersection table of (students' educational level * personalization (personal closet)).*

By examining the table above, we find that this need is more pronounced in educationally poor and educationally moderate students than in students with high educational status. That is, as many as 54 moderate students and 56 poorly educated students welcomed the idea of having personal lockers, while this number is only 33 in students with high educational status. This indicates the importance of the issue of creating individual identity in students who are not in a good educational situation. The frequency chart is shown in **Figure 7**.

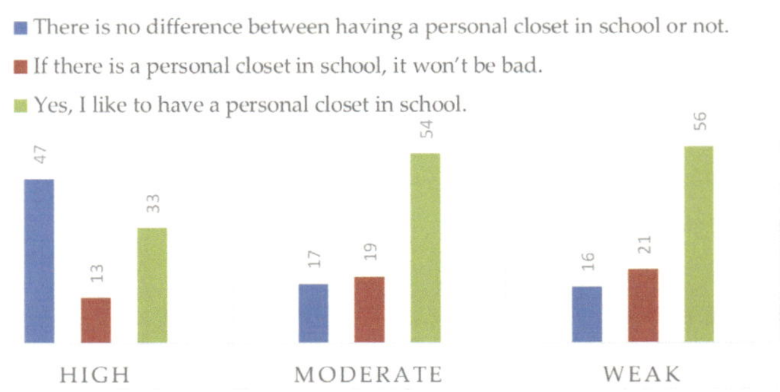

Figure 7.
The frequency of the first hypothesis.

2. Analysis of the second hypothesis: "The lack of favorable environmental conditions in the classroom (the comfort temperature and appropriate light) reduces the interest in participating in class activities and negatively impacts on educational efficiency"

This hypothesis is based on the inadequacy of classroom environment conditions and examines the impact on student learning. The mentioned hypothesis is evaluated by asking this question to the students: "Have you ever experienced falling asleep or dislike listening to teacher because of the uncomfortable temperature or light of your class?"

The test used for this hypothesis is a single-variable Chi-square test whose results are shown in **Table 6** as follows:

Students' educational level		Functional requirements			Total
		Yes, it has happened.	It has happened sometimes.	It has never happened.	
High	Count	13	7	73	93
	Expected count	11.1	19.2	62.7	93.0
Moderate	Count	9	31	50	90
	Expected count	10.8	18.6	60.7	90.0
Weak	Count	12	24	57	93
	Expected count	11.1	19.2	62.7	93.0
Total	Count	34	62	181	276
	Expected count	33.0	57.0	186.0	276.0

Table 6.
*Intersection table of (students' educational level * functional requirements).*

Table 6 above shows that there is no problem in the classes in terms of comfortable temperature from the viewpoints of students. This issue is answered uniformly in all three groups of students, and this hypothesis is accordingly rejected. The frequency chart is shown in **Figure 8**.

3. Analysis of the third hypothesis: "The availability of cozy spaces and offshoots outside the classroom to study or perform personal activities of students increases their sense of attachment to the school"

One of the patterns for personalizing the design of schools is the design of spaces of varying quality for small groups or individual activities. Therefore, the impact of these cozy spaces and quiet places on students' sense of attachment to the school is

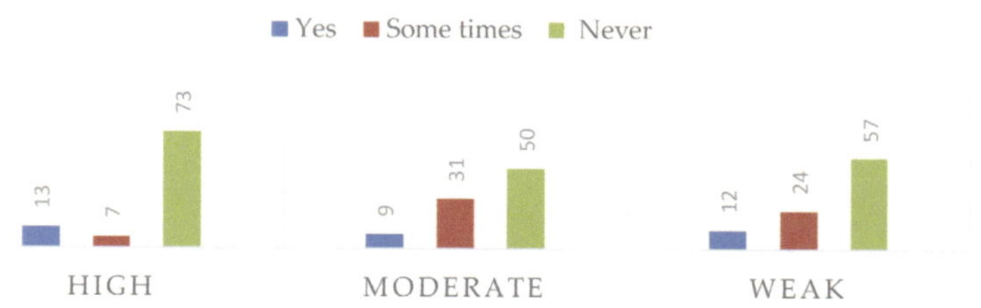

Figure 8.
The frequency of the second hypothesis.

Figure 9.
A cozy and peaceful area of school, where students can study or perform personal activities, designed by authors.

analyzed. An example of such spaces is shown in **Figure 9**. The mentioned hypothesis is evaluated by asking this question to the students: "Will you like your school more if there are cozy and quiet places and you can do your favorite activities?"

The test used is a single-variable Chi-squire whose results are shown in **Table 7** as follows:

Students' educational level		Functional requirements			Total
		Yes, it has happened.	It has happened sometimes.	It has never happened.	
High	Count	13	7	73	93
	Expected count	11.1	19.2	62.7	93.0
Moderate	Count	9	31	50	90
	Expected count	10.8	18.6	60.7	90.0
Weak	Count	12	24	57	93
	Expected count	11.1	19.2	62.7	93.0
Total	Count	34	62	181	276
	Expected count	33.0	57.0	186.0	276.0

Table 7.
*Intersection table of (students' educational level * personalization (personal space)).*

The results in **Table 7** show that none of the three student groups are interested in having such spaces in the school. A total of 168 out of 276 people reported the ineffectiveness of such spaces in their degree of interest to the school. The frequency chart is shown in **Figure 10**.

Figure 10.
The frequency of the third hypothesis.

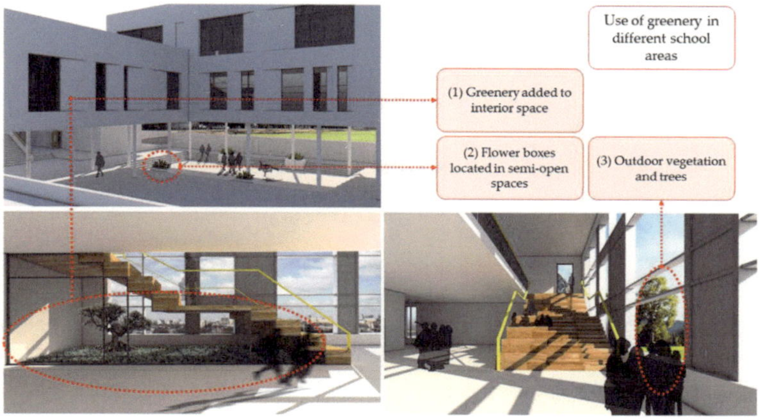

Figure 11.
Presence of green spaces in the school (in the school yard, inside the building, and in classes), designed by authors.

4. Analysis of the fourth hypothesis: "The presence of green spaces in the school (in the schoolyard, inside the building, and in classes) can increase the sense of attachment to the school." A few examples are shown in **Figure 11**.

In many studies, the undeniable role of interacting with the natural environment has been mentioned in the sense of attachment to the place. To this end, this issue was investigated. For evaluating this hypothesis, the students were asked to answer the following question: "Will you feel better about school if several trees are planted in the school yard or some greenery are added to some places?"

The statistical method used is a single-variable Chi-square model; the results of which are shown in **Table 8** as follows:

Students' educational level		Natural elements (greenery)			Total
		No	Yes	Yes, a lot	
High	Count	8	22	63	93
	Expected count	13.1	17.2	62.7	93.0
Moderate	Count	11	13	66	90
	Expected count	12.7	16.6	60.7	90.0
Weak	Count	20	16	57	93
	Expected count	13.1	17.2	62.7	93.0
Total	Count	39	51	186	276
	Expected count	39.0	51.0	186.0	276.0

Table 8.
*Intersection table of (students' educational level * natural elements and (greenery)).*

The results in **Table 8** point to the great importance of this factor in creating the sense of place attachment. Of the 276 participants, a total of 237 students expressed their interest in having vegetation in the school environment.

The frequency diagram of the relationship between the presence of vegetation and the sense of attachment is shown in **Figure 12**.

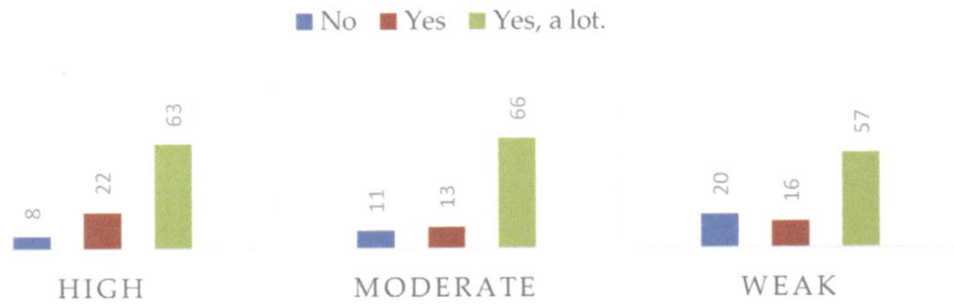

Figure 12.
The frequency of the fourth hypothesis.

5. Analyzing the fifth hypothesis: "Open spaces within the school designed for students' gathering and engaging in group activities can increase interpersonal interaction and increase the sense of attachment to the school"

The factors that led to increased personal identities and a sense of social and physical attachment are the possibility of social interactions that are provided by the environment. This section analyzes the impact of this factor on students' sense of attachment to the school. The mentioned hypothesis is evaluated by asking this question to the students: "Will you feel better about school if the school has a spacious hall, where you can spend time with your friends instead of going to the school yard or class?"

The method used to measure this hypothesis is the single-variable Chi-square test; the results of which are presented in **Table 9** as below:

Students' educational level		**Interaction with others**			**Total**
		I prefer the yard.	**I have not thought about it yet.**	**Yes, I like such a place.**	
High	Count	19	35	39	93
	Expected count	18.2	28.3	46.5	93.0
Moderate	Count	11	22	57	90
	Expected count	17.6	27.4	45.0	90.0
Weak	Count	24	27	42	93
	Expected count	18.2	28.3	46.5	93.0
Total	Count	54	84	138	276
	Expected count	54.0	84.0	138.0	276.0

Table 9.
*Intersection table of (students' educational level * interacting with others).*

The results in **Table 9** illustrate that interested students have spaces where they can spend more time with their friends, separate from the usual schoolyard pattern. The frequency chart is shown in **Figure 13**.

6. Analysis of the sixth hypothesis: "Classes that have larger windows with a suitable perspective are more appealing to students"

The question for evaluating this hypothesis is: "Will you like your class more if it has larger windows with a suitable perspective?"

Table 10 shows the impact of the proper outlook of the class and its positive impact on student interest in the class.

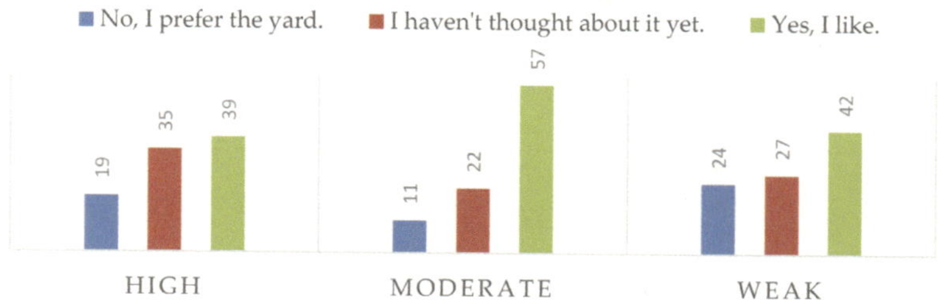

Figure 13.
The frequency of the fifth hypothesis.

Students' educational level		Perspective and communication with the outside			Total
		No, I do not like.	**There is no difference.**	**Yes, I like.**	
High	Count	4	44	45	93
	Expected count	4.0	38.4	50.5	93.0
Moderate	Count	5	40	45	90
	Expected count	3.9	37.2	48.9	90.0
Weak	Count	3	29	61	93
	Expected count	4.0	38.4	50.5	93.0
Total	Count	12	113	151	276
	Expected count	12.0	114.0	150.0	276.0

Table 10.
*Intersection table of (students' educational level * perspective and communication with the outside).*

A very limited number of students were uninterested in having a suitable perspective (a total of 12); 150 were positive about this. Also, the number of students with interest in this topic is significant (60 people). The frequency chart is shown in **Figure 14**.

7. Analysis of the seventh hypothesis: "The lack of clarity in the design of the class, the closure of its space, and the lack of visual communication with the outside create unpleasant feelings among students"

The question for evaluating this hypothesis is: "Do you like your class to be more spacious and not to be surrounded by walls from every side?"

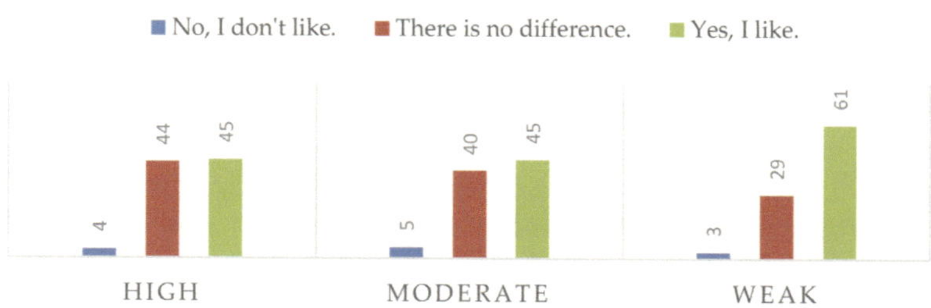

Figure 14.
The frequency of the sixth hypothesis.

Students' educational level		Transparency in design			Total
		Yes, a lot.	If it is, it will not be bad.	I like the current class.	
High	Count	31	30	32	93
	Expected count	34.4	25.3	33.4	93.0
Moderate	Count	30	20	40	90
	Expected count	33.3	24.5	32.3	90.0
Weak	Count	47	25	21	93
	Expected count	34.4	25.3	33.4	93.0
Total	Count	108	75	93	276
	Expected count	102.0	75.0	99.0	276.0

Table 11.
*Intersection table of (students' educational level * transparency in design).*

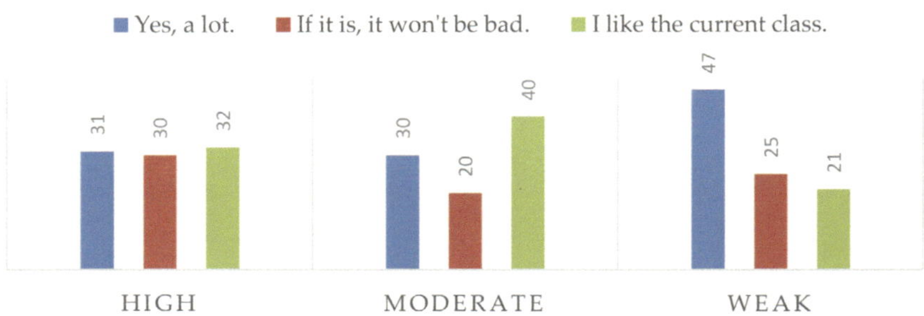

Figure 15.
The frequency of the seventh hypothesis.

In **Table 11** below, the effect of the external and internal visual impact on the sense of attachment of the students has been investigated.

The degree of interest in more open spaces and visual visibility is inversely correlated with the educational level. The other way around is the case with low achiever students. The frequency chart is shown in **Figure 15**.

8. Analysis of the main hypothesis: "The students' sense of attachment to school has a positive impact on their educational efficiency"

In this section, the main hypothesis of the research is analyzed (the relationship between students' sense of attachment to the school and their educational efficiency is examined).

Intersection **Table 12** presents the results as follows:

The frequency chart is shown in **Figure 16**.

As it is shown, there is a direct relationship between the students' educational efficiency and their sense of attachment to the school. In students with excellent educational status, there is significant interest in the school, while this has decreased in low achiever students educationally, and almost half of these students either do not like the school at all or they have little interest in school. In fact, the sense of attachment factor can be considered as one of the factors that influences educational efficiency. At the same time, it is also necessary that the sense of attachment is not the only factor affecting the level of educational outcome, but can

Students' educational level		Sense of attachment to school			Total
		Not at all	A little	A lot	
High	Count	4	20	69	93
	Expected count	3.0	30.3	59.6	93.0
Moderate	Count	5	25	60	90
	Expected count	2.9	29.3	57.7	90.0
Weak	Count	10	48	35	93
	Expected count	3.0	30.3	59.6	93.0
Total	Count	19	93	164	276
	Expected count	9.0	90.0	177.0	276.0

Table 12.
*Intersection table of (students' educational level * sense of attachment to the school).*

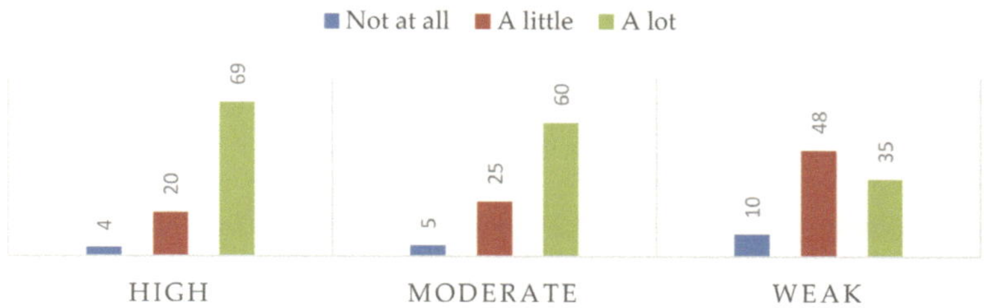

Figure 16.
The frequency of the main hypothesis.

be increased by creating an environment with which students have an affinity. This in turn has a positive effect on their interest in the school environment and learning.

Based on what has been done so far in the topic of effects of place attachment on educational efficiency in schools, there are opportunities to expand the research on the following directions:

- Implementing sense of place and identity for cultural sustainability [57]
- Benefits of schools' attachment and sense of identity among children for smart cities [58]
- Issues and opportunities for sustainable development of schools to sustainable development of societies [59]
- The effect of providing the approaches to increase sense of attachment among students on sustainability in higher education [60]
- Developing a unified framework for academic research on sustainable school attachment and smart city concept [61].

5. Conclusion

Recently, there is a significant increase toward understanding the importance of education in developing countries. One of the factors that plays a vital role in educational development is known as sense of attachment to the schools' environment. The main aim of this research was to investigate effective factors to enhance sense of attachment among students in schools' environment through design. Then, to justify the effectiveness of these factors, the positive impacts of creating sense of attachment in students were presented. Thus, in this research, a number of questions and hypothesis are developed. Then, a questionnaire was designed to demonstrate the effectiveness of factors and to validate them. The survey targeted 278 students of 10 state primary schools of Shahriyar, Iran. The final findings analyzed by SPSS indicate that there is a meaningful relation between students' educational efficiency and their sense of attachment to school. Data analysis also shows that important factors, such as social activities and connection with natural elements in design, can have a noticeable impact on the students' sense of attachment to school as well as their educational efficiency. The results of the research carried out are as follows:

1. Primary school students are more interested in social issues and engaging with others than personal issues (cozy places and personal closets).

2. Transparency in the design and communication between the inside and outside of the different parts of the school building will make students more interested in and feel more attached to the school.

3. Communicating with the natural environment, providing open and green spaces in the school, and directly connecting these spaces with students represent a very important factor in increasing pupils' sense of belonging to the school.

4. Students who are low achievers educationally are more inclined to gravitate toward other topics, such as being transparent in designing and communicating with the natural environment and using trees; thus, it seems that these aspects can have more of an impact on them than stronger students.

5. Educational quality and outcomes in students are affected by their sense of attachment to the school, so that low achiever students are less interested and stronger students are more interested in school.

Although, in this research, most important factors were investigated through analyzing the students' responses, there are still some fundamental aspects of this subject that need to be covered in future research. The link between these factors to the cultural and social background of students with the design principles of the school is a broad subject that can be extended in future studies.

Author details

Farhad Soheili[1], Reyhaneh Karimi[1]*, Behnaz Avazpour[2] and
Samad M.E. Sepasgozar[2]

1 University of Science and Culture, Tehran, Iran

2 University of New South Wales, Sydney, Australia

*Address all correspondence to: reyhanehkarimi1992@gmail.com

References

[1] Maxwell LE, Chmielewski EJ. Environmental personalization and elementary school children's self-esteem. Journal of Environmental Psychology. 2008;**28**(2):143-153

[2] Dewey J. The School and Society and the Child and the Curriculum. USA: University of Chicago Press; 2013

[3] Burke C, Grosvenor I. The School I'd Like: Children and Young People's Reflections on an Education for the 21st Century. London, UK: Routledge; 2003

[4] Edwards CP, Gandini L, Forman GE. The Hundred Languages of Children: The Reggio Emilia Approach–Advanced Reflections. California, USA: Greenwood Publishing Group; 1998

[5] Bradley WS. Expecting the Most from School Design. Education Resources Information Center (ERIC). 2000

[6] Christopher G. Effect of architecture on education. Educational Facility Planner. 1991;**29**(1):11-15

[7] Lipsitz J. Successful Schools for Young Adolescents. New York, USA: Routledge; 2019

[8] Brookover WB, Schweitzer JH, Schneider JM, Beady CH, Flood PK, Wisenbaker JM. Elementary school social climate and school achievement. American Educational Research Journal. 1978;**15**(2):301-318

[9] Weber S, Kronberger N, Appel M. Immigrant students' educational trajectories: The influence of cultural identity and stereotype threat. Self and Identity. 2018;**17**(2):211-235

[10] Ma X. Sense of belonging to school: Can schools make a difference? The Journal of Educational Research. 2003; **96**(6):340-349

[11] Killeen JP, Evans GW, Danko S. The role of permanent student artwork in students' sense of ownership in an elementary school. Environment and Behavior. 2003;**35**(2):250-263

[12] Upitis R. Four strong schools: Developing a sense of place through school architecture. International Journal of Education & the Arts. 2007;**8**:1-16

[13] Earthman GI. Varia: The quality of school buildings, student achievement, and student behavior. Bildung und Erziehung. 1999;**52**(3):353-372

[14] Moore GT. The designed environment and cognitive development: A brief review of five domains of research. Children's Environments Quarterly. 1985;**2**(2):26-33

[15] Moore GT. Ready to learn: Toward design standards for child care facilities. Educational Facility Planner. 1994; **32**(1):4-10

[16] Ahrentzen S. In: Evans GW, editor. Environmental Stress. New York: Cambridge University Press; 1982

[17] Knight G, Noyes J. Children's behaviour and the design of school furniture. Ergonomics. 1999;**42**(5): 747-760

[18] Ehrenberg RG, Brewer DJ, Gamoran A, Willms JD. Class size and student achievement. Psychological Science in the Public Interest. 2001;**2**(1): 1-30

[19] Evans GW, Maxwell L. Chronic noise exposure and reading deficits: The mediating effects of language acquisition. Environment and Behavior. 1997;**29**(5):638-656

[20] Maxwell LE. Multiple effects of home and day care crowding.

Environment and Behavior. 1996;**28**(4): 494-511

[21] Maxwell LE. Home and school density effects on elementary school children: The role of spatial density. Environment and Behavior. 2003;**35**(4): 566-578

[22] Tanner CK. Effects of school design on student outcomes. Journal of Educational Administration. 2009; **47**(3):381-399

[23] Walberg HJ. Improving Educational Standards and Productivity: The Research Basis for Policy. McCutchan Publishing Corporation; 1982. pp. 237-288

[24] Phillips R. Educational facility age and the academic achievement of upper elementary school students [unpublished doctoral dissertation]. Georgia: University of Georgia; 1997

[25] Finn JD, Achilles CM. Tennessee's class size study: Findings, implications, misconceptions. Educational Evaluation and Policy Analysis. 1999; **21**(2):97-109

[26] Duke DL. Does it Matter where our Children Learn? Education Resources Information Center (ERIC). 1998

[27] Cotton K. School Size, School Climate, and Student Performance. Portland, OR: Northwest Regional Education Laboratory; 1996

[28] Stevenson KR. Elementary school student capacity: What size is the right size? Educational Facility Planner. 1996; **33**(4):10-14

[29] Falk JH, Dierking LD. School field trips: Assessing their long-term impact. Curator: The Museum Journal. 1997; **40**(3):211-218

[30] Eaton D. Cognitive and Affective Learning in Outdoor Education. Canada: National Library of Canada=Bibliothèque Nationale du Canada; 2000

[31] Ballantyne R, Packer J. Nature-based excursions: School students' perceptions of learning in natural environments. International Research in Geographical and Environmental Education. 2002;**11**(3):218-236

[32] Mozafar F, Mahdi Zade Seraj F, Mir Moradi S. Recognizing the role of nature in educational spaces. Journal of Education Technology. 2009;**1**:37-46

[33] Kamel NH. Learning Environment Design Grammar. Tehran, Iran: Sobhane Noor; 2009

[34] Sami AA. The concept and function of open space in traditional and new schools. Soffeh. 2000;**10**:104-111

[35] Seddigh Z, Dulabi P, Karimi B. Space perception; a key element in designing children's spaces. In: Fourth International Conference on Civil Engineering, Architecture and Urban Economics Development; Shiraz, Iran. 2017

[36] Mardomi K, Delshad M. Flexible learning environment (an experienced child's world, a changeable educational system). Journal of Iranian Architecture and Urbanism. 2010;**1**:109-118

[37] Zarghami E, Ghanbaran A, Pahlavani M. Criteria for designing future education space. In: The First National Education Conference in Iran 1404; Tehran, Iran. 2011

[38] Roghani M. Study of the environmental components of learning space and its impact on student behavior. In: The First Scientific Congress of Modern Horizons in Architecture, Civil Engineering, Culture and Urban Management of Iran; Tehran, Iran. 2015

[39] Mehdi Nezhad J, Zarghami E, Shariatinia F. A study of the way to

create a sense of place for restoration of identity and confidentiality in architectural areas. In: The Third International Conference of Applied Research in Civil Engineering, Architecture and Urban Management; Tehran, Iran. 2015

[40] Zare M. The effect of school architecture on the students' learning process. In: Third Scientific Conference on Modern Horizons in Geography and Urban Planning in Iran. 2016

[41] Golgavand F. Comparison of the effect of current teaching method with teaching method based on research and exploration in the first grade secondary school. Curriculum Research. 2007: 95-115

[42] Hidalgo MC, Hernandez B. Place attachment: Conceptual and empirical questions. Journal of Environmental Psychology. 2001;**21**(3):273-281

[43] Shumaker SA, Taylor RB. Toward a clarification of people-place relationships: A model of attachment to place. Environmental psychology: Directions and Perspectives. 1983;**2**:19-25

[44] Hummon DM. Community Attachment. Place Attachment. New York, USA: Springer; 1992. pp. 253-278

[45] Rubinstein N, editor. There's no place like home: home as trauma: lessons of the unspoken. In: Power by Design: The Proceedings of the Twenty-Fourth Annual Conference of the Environmental Design Research Association. Oklahoma: EDRA Press; 1993

[46] Parker R, Levinson MP. Student behaviour, motivation and the potential of attachment-aware schools to redefine the landscape. British Educational Research Journal. 2018;**44**(5):875-896

[47] Amitay G, Rahav G. Attachment and pedagogical relevant practices as elements of a successful alternative school through the narratives of its students. Psychology in the Schools. 2018;**55**(10):1239-1258

[48] Javan Forouzande A, Matlabi G. The sense of belonging to the place and its constituent elements. Journal of City Identity. 2011;**8**:27-37

[49] Lynch K. The Image of the City. Cambridge, UK: MIT Press; 1960

[50] Lewicka M, Rowiński K, Iwańczak B, Bałaj B, Kula AM, Oleksy T, et al. On the essentialism of places: Between conservative and progressive meanings. Journal of Environmental Psychology. 2019;**65**:101318

[51] Dazkir SS. Place meaning, sense of belonging, and personalization among university students in Turkey. Family and Consumer Sciences Research Journal. 2018;**46**(3):252-266

[52] Rogoff B. Cognition as a Collaborative Process. Handbook of Child Psychology, Cognition, Perception, and Language. 1998

[53] Riger S, Lavrakas PJ. Community ties: Patterns of attachment and social interaction in urban neighborhoods. American Journal of Community Psychology. 1981;**9**(1):55-66

[54] Mokhtarmanesh S, Ghomeishi M. Participatory design for a sustainable environment: Integrating school design using students' preferences. Sustainable Cities and Society. 2019;**51**:101762

[55] Ryan RL. Exploring the Effects of Environmental Experience on Attachment to Urban Natural Areas. Environment and Behavior. January 2005;**37**(1):3-42. DOI: 10.1177/0013916504264147

[56] Nair P, Fielding R. The Language of School Design: Design Patterns for 21st Century Schools. 2nd ed. Washington, USA: Designshare Inc.; 2009

[57] Semken S, Brandt E. Implications of sense of place and place-based education for ecological integrity and cultural sustainability in diverse places. In: Cultural studies and environmentalism. New York, USA: Springer; 2010. pp. 287-302

[58] Belanche D, Casaló LV, Orús C. City attachment and use of urban services: Benefits for smart cities. Cities. 2016;**50**: 75-81

[59] Dempsey N, Bramley G, Power S, Brown C. The social dimension of sustainable development: Defining urban social sustainability. Sustainable Development. 2011;**19**(5):289-300

[60] Barlett PF, Anderson E, Boyer JC, Brunckhorst D, Princen T, Barlett PB. Reason and reenchantment in cultural change: Sustainability in higher education. Current Anthropology. 2008; **49**(6):1077-1098

[61] Furlong MJ, Whipple AD, Jean GS, Simental J, Soliz A, Punthuna S. Multiple contexts of school engagement: Moving toward a unifying framework for educational research and practice. The California School Psychologist. 2003;**8**(1):99-113

Chapter 6

Effective Factors on Desirability of Private Open Spaces: A Case Study of Kuye Nasr Residential Buildings, Tehran

Reyhaneh Karimi, Behnaz Avazpour and Samad M.E. Sepasgozar

Abstract

The gradual transformation of courtyard houses to apartments has contributed to the omission of certain spaces. The ill-matching of modern housing, with people's lifestyles, has caused undesirable changes to their quality of life, health and well-being. Providing houses that suit individuals' lifestyles plays an important role in the building social sustainability as well as economic and environmental aspects; this article will investigate, alongside, ways to improve private open areas in apartments. Two- to six-floor residential buildings in Kuye Nasr neighborhood were chosen to find factors that can improve the desirability of private open areas. First, a number of factors were collected by taking photos and conducting field studies. Next, several residents were randomly asked to share their opinions about such areas. Finally, a questionnaire examined the reliability of factors, which was then distributed to 100 residents. The final data was analyzed utilizing SPSS23. This study reveals that the functional aspect of a balcony is the most important item among semantic, perceptual, functional, physical, environmental, and beauty aspects. The following factors can contribute to residents' satisfaction with balconies: dimensions; use of plants; connection with sky; peace and comfort; safety; function; and beauty. Findings additionally suggests the connection of balcony with kitchen.

Keywords: desirability factors, private open area, balcony, residential building, Kuye Nasr neighborhood

1. Introduction

The insufficiency and high price of land have dramatically increased apartment living and as well as people's willingness [1, 2] to settle in residential units over recent decades [3]. The gradual conversion of courtyard houses to apartments and residential units has led to changes in the function or removal of spaces embedded in Iranian lives [3], illustrated in **Figure 1**. The vacuum created by the incompatibility of new housing—a significant part of which includes residential complexes and buildings—has caused unfavorable changes in human life [4, 5]. The land price is a major issue that limits individuals' choices, forcing them to buy units of a small size. Due to land prices, builders and designers prefer to eliminate open and semi-open spaces to increase and

IntechOpen

replace indoor spaces [3, 6, 7]. Thus, spaces such as indoor and outdoor communication spaces (e.g., balconies) are rarely seen, and the removal of such important spaces from human life is more prevalent nowadays. This modern norm overlooks the necessity of developing housing that conforms to the lifestyle of its residents. What's more, the number of houses that can satisfy all the needs of Iranian families is too restricted. Thus, this research aims to investigate the removal or negligence of private open spaces that would ordinarily bring enjoyment to residents. Furthermore, this study seeks to share insights that will improve the quality of people's life environments and ultimately help them benefit more from their home spaces. The research background indicates a gap on how to design balconies that satisfy the residents of residential buildings. A scientific study into this area could be utilized in future projects to help enhance the quality of such home environments (**Figures 1–5**).

For example, Einifar and Ghazizadeh [10] showed that designing outdoor spaces is less dealt with by designers; rather, their main focus being the interior space of the building, without paying enough attention to the spaces between the outdoor and indoor space. The authors discuss the necessity and importance of designing the lost and forgotten space of buildings like open space, and recommend that designers of residential complexes pay more attention to such spaces.

Extant studies on private open spaces consider a range of views on residential building spaces including behavioral, environmental, physical, perceptual, semantic, and esthetics. What sets the current research apart from past studies into this subject is its seeking which one of these factors is more influential in this statistical population.

Designers are obliged to see all the approaches in designing open spaces while at the same time pay close attention to the aspects regarded as important by the residents. One article [11] examined the three categories of perceptual-semantic, functional-behavioral and physical-environmental factors in a social-cultural context, with the balcony selected as a private open space between the residential unit and the residential area. The main question of this research was what factors affect the quality of the balcony as an intermediate space and what do people expect from its function in different socio-cultural backgrounds? Based on interview, they examined three residential complexes and presented a model containing six factors. The authors concluded that attention to the secondary functions and the multipurpose nature of intermediate spaces should be among the primary concerns of planning and designing to enhance the quality of life in apartment housing.

Badeie [12], on the other hand, expressed her research findings from a philosophical point of view: wall space-separators are not necessarily the only effective separation and limitation, but the space instruction of walls is characterized by

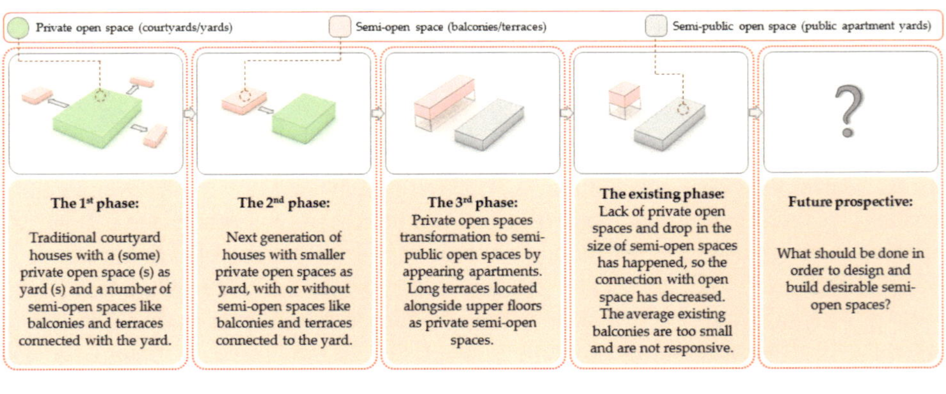

Figure 1.
Gradual transformation of Iranian traditional courtyard houses to new apartments.

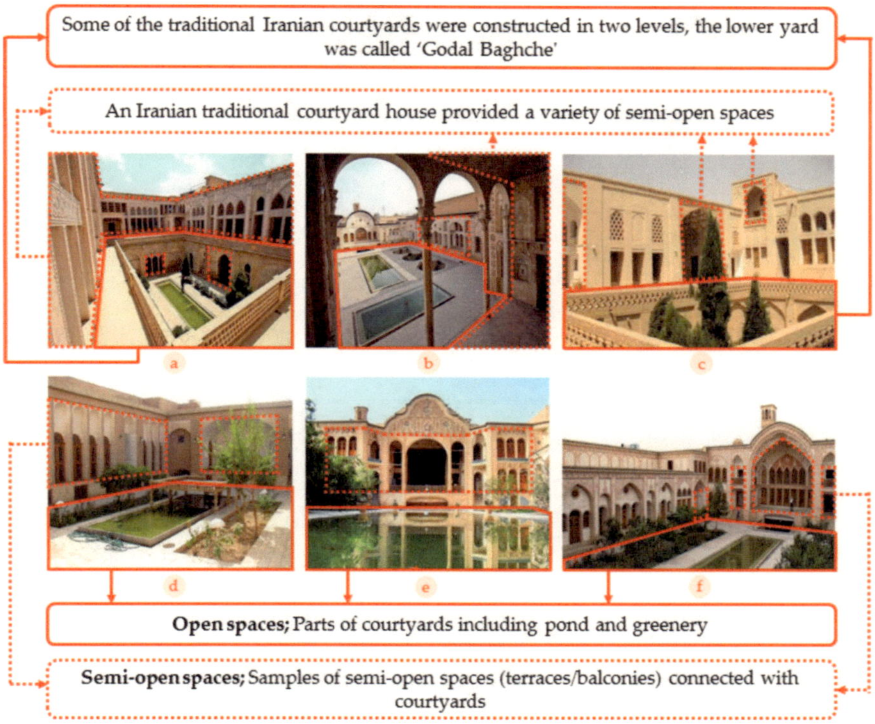

Figure 2.
Samples of Iranian traditional courtyard houses: (a) Abbasi House, Kashan, (b) Tabatabaei House, Kashan, (c) Pirnia House, Yazd, (d) Navab Vakil House, Yazd, (e) Borujerdis House, Kashan, and (f) Ameri House, Kashan; adopted from [8], analyzed by authors.

Figure 3.
The second generation of Iranian houses with yard, adopted from [9], analyzed by authors.

inferential definition of connectivity. Privacy creates protection, security, and confidentiality in the design of appropriate space frame of artifact environment on one hand and fortifies the dynamic and diverse capabilities of space components

Figure 4.
Old apartments designed with terraces or balconies as private open spaces [ref: photos are taken and analyzed by researchers, 2019].

Figure 5.
Younger than 5-year-old apartments designed with balconies as private open spaces [ref: photos are taken and analyzed by researchers, 2019].

for coherence on the other hand. The research focused on enhancing the capabilities of walls in creating links and distance; the capabilities resulting from the concepts of privacy and connection enriched the walls' functionality area in the construction of space (**Figure 6**).

Another research [13] emphasizes the need for intermediate spaces, asserting that joint spaces like balconies play the roles of both interior and exterior spaces and are regarded as the distance between outside and inside space. Also, Mir Shahzadeh [14] examined the role of the linking boundary space, such as balcony and porch, etc., in producing meaning.

Christopher Alexander reported a British study in the Architects Magazine (1957) concerning the balconies of apartments and small houses. He stated that two-thirds of people, for their own reasons, have never used balconies that lack privacy. And in "a pattern language" [15], a model is presented for balconies according to physical characteristics; yet only the quantitative aspect is considered in the book—for example, a minimum depth of 180 cm is considered for such spaces.

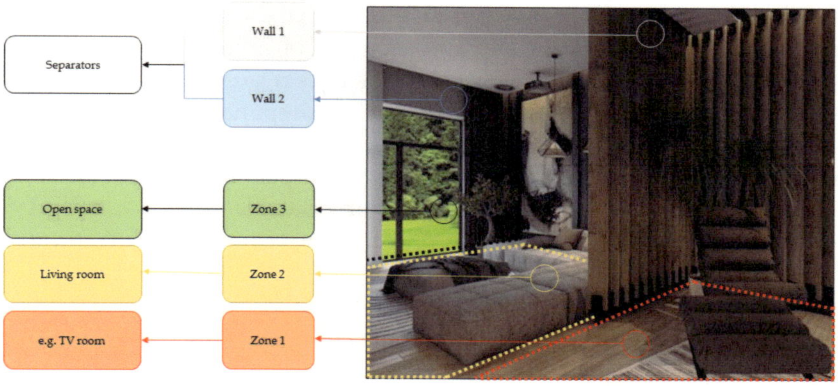

Figure 6.
Use of walls as space separators, by which links and distances can be made, adopted from [9] and analyzed by authors.

Pierre Bourdieu, a well-known French sociologist and anthropologist, also acknowledges that different lifestyles will create a distinction in the living space; the higher the social hierarchy, the higher the esthetic properties of space than its functional responses will be. Therefore, we will have distinct spaces based on necessity-favored need or luxury-favored need, an issue that can be reviewed in the balconies, but not in the present article [16].

Research by Mazandarani [17] investigating housing developments from past to present represents the foundations and status of the intermediate space. The importance of the spaces discussed in the article is underlined by the authors, and general rulings on the points that should be considered by the designers are presented. Unfortunately, there is no comprehensive discussion on balconies as discussed in article [17].

Among other international books in this field is the Book of Life between Buildings, which deals with the concepts of public open spaces, and the initial principles that can be used for intermediate spaces [18]. In the book Esthetics in Architecture [19], regarding the interactive spaces of interior and exterior, the author divides such spaces into three categories: (1) architecture without interaction with outside; (2) architecture with interaction with outside; and (3) modern architecture and border distortions.

In another study, in the chapter on zones and thresholds, the importance of focusing on the scopes that provide a correct definition of outside and inside space for the user of environment is presented. It is argued that humans create limits to understand that we belong within this environment and that we are safe. Furthermore, thresholds and passage spaces are places where the environment manifests itself. Stairs, the edges of roofs, gates, doors, balconies, windows, etc. are all regulators of this manifestation and control the penetrability of these borders. These design factors approve the space separation while at the same time create the possibility of physical and visual passages for people. Hence, the author expresses the necessity and importance of addressing such spaces, and expresses the role of such spaces in three parts: (1) use-oriented role (functional); (2) protective role, or a controlled space, through which the perspective can be seen. This role can be divided into two parts, social and physical; (3) the semantic role, being that the signs of each place, according to the common conventions and social traditions, have a special behavior in each corner and within the desired range. About the balconies, the author says: "Balconies are not practically considered as privacy zone and play the role of seeing and being seen more than the terrace and porches" [20].

Another book on this subject is the work of Rudolf Arnheim, which states: no spatial issue is as much an inherent characteristic of an architect's work given that they should consider the inside and outside spaces as related concepts. That is, indoor and outdoor areas should be considered as components of one concept [21].

To summarize, **Table 1** includes all the related literature.

Researchers looked at these spaces from different perspectives. But, as it is clear to every researcher, architecture is not meaningful outside of its context and should be designed and developed on the basis of its main context and social, cultural, and physical conditions.

Therefore, this study tries to help architects to define and explode a specific context in order to have optimal design of open spaces in residential buildings of a neighborhood. By using factors that have been previously investigated and extracted by previous studies, this research uses the case study the Kuye Nasr neighborhood to examine the lifestyle of the residents of the residential buildings. It considers the physical conditions of the buildings, the instructions and rules for improving their design quality, and the desirability of open spaces within residential buildings (balconies). The abovementioned factors will be then evaluated and the results will be available to designers as executable and functional factors.

It should be noted that during the research process, it was found out that the residents of residential buildings in Kuye Nasr were more concerned about the

Author or researcher	Year	Subject or title	Theoretical perspective	Approach
Swapan, et al. [22]	2019	Understanding the importance of front yard accessibility for community building: a case study of Subiaco, Western Australia	Front yards as semi-public-private areas can play an important role in the residents' sense of community	Behavioral
Reuben, et al. [23]	2019	Residential neighborhood greenery and children's cognitive development	The whole study puts an emphasis on the contribution of surrounding open space greenery of residential buildings on the cognitive development on the children, who were raised in such places.	behavioral, environmental, esthetic
Azad, et al. [24]	2018	Effect of housing layout and open space morphology on residential environments—applying new density indices for evaluation of residential areas case study: Tehran, Iran	Putting an emphasis on the vitality of presence of private or public open spaces and the importance of considering them in the construction laws.	Physical-environmental
Milanović and Vasilevska [2]	2018	Influence of private open spaces on the quality of living in low-rise high density housing	The research focuses on the advantages of private open areas	Behavioral physical
Einifar and Ghazizadeh [10]	2010	Typology of residential complexes of Tehran with open space measure	Importance of paying attention to lost spaces such as open spaces in residential buildings	Physical, behavioral

Author or researcher	Year	Subject or title	Theoretical perspective	Approach
Einifar and Ali Niaye Motlagh[11]	2014	Explaining the concept of outside-inside in in-between spaces of apartment housing "the case study of balcony in three types of Tehran residential complexes	Considering the secondary functions and multi-purpose nature, the intermediate spaces are among the planning, and design needs to improve the quality of life in apartment housing	Physical-environmental
Badeie [12]	2002	Walls-privacy connection	Philosophical review of connection space	Meaning, behavioral
Rezakhani [13]	2013	An introduction to the concept of joint in architecture based on Heidegger methodology of word etymology	Joint etymology	Meaning
Mirshahzadeh, et al. [14]	2013	The role of boundary space in the creation of meaning	Semiotics approach in boundary-connection space	Esthetic, meaning
Alexander [15]	1977	A pattern language: towns, buildings, construction	Physical examination	Physical
Bourdieu [16]	2013	Distinction: A social critique of the judgment of taste.	Effect of lifestyle on spaces	Behavioral
Haeri Mazandarani [17]	2009	Home, culture and nature	Past developments have changed for the present day and represent the foundations of the space between them	Physical
Gehl [18]	2011	Life between buildings: using public space	Expression of the rules of public open space	Behavioral, physical
Grütter [19]	2014	Grundlagen der Architektur-Wahrnehmung	The division of architectural time periods regarding the relationship between inside and outside space	Physical, perceptional
Von Mayes [20]	2013	Elements of architecture: from form to place	The importance of focusing on the scopes as the elements separating inside and outside	Behavioral, physical
Rudolf Arnheim [21]	1977	The dynamics of architectural form	Relation between inside and outside and attention to this in architecture	Meaning, behavioral
Bentley [25]	1985	Responsive environments: a manual for designers	The strong role of designers in designing flexible ways to keep track of the budget	Behavioral, meaning
Lawson [26]	2007	Language of space	It addresses the issue of realm and perceptual mechanisms in space	Behavioral, meaning

Table 1.
Selected papers referring to open spaces and relevant measures.

functional aspect of balconies. Hence, this component has been emphasized more than several other factors influencing the usefulness of balconies. The main questions that this research tackles are what factors can satisfy the residents of the Kuye Nasr residential buildings regarding their balconies, and what is the association between the influential factors in the desirability of private open spaces?

To achieve this, the concerns, needs, and opinions of the residents of Kuye Nasr were determined through a field study and interviews, based on which a number of factors were extracted. The validity of such factors was then examined by means of a questionnaire; those factors with a higher statistical weight were presented as features that should be applied by designers of balconies.

2. Theoretical framework

2.1 The connection between interior and exterior space

The interior of each place is distinguished from the exterior and the surrounding environment by various spatial elements. People have always lived in both outside and inside environments and must be active between these two poles. Consequently, both of these currents cannot be separated, but it is possible that the impact of one of them outweighs the impact of the other. In fact, the kind of governing relationship between the external and internal environment is determined by the spatial relationships between these elements. The structuring and arrangement of these relationships (relations between inside and outside) result from the contradiction between them; that is to say, the separation of a protective shelter from the surrounding area on one hand, and the establishment of the relationship between the two locations on the other hand, are essential for human life. This contradiction is revealed only when the characteristics of the inside and outside location are compared. In spite of these contradictions, there are relations between the inside and outside. According to Christian Norberg Schulz, the relations and conditions governing these two create the art of architecture [19].

Many activities take place around low-rise residential buildings with direct access to the outside. What's more, there is a flow between the inside and the outside; for instance, in order to know what is going on outside, residents can quickly step out or drink a cup of coffee on the stairs. While multi-story buildings, have less inhibiting them from stepping outside and being among their residential community [18].

Therefore, the role and influence of the intermediaries of the outer and inner space (such as the balcony) are not only undeniable, but also quite significant. Hence, by eliminating or ignoring each of them, the interfacing of the inner and outer spaces is disturbed and the quality of spatial sequencing is declined. Therefore, maintaining the identity of such spaces and optimizing their architectural design is of great importance today.

2.2 Private open spaces

The border between the closed space and the surrounding open space can be understood as the interface of two worlds. The set of elements—the main function of which is to keep the open space from being enclosed— is called a building and plays a significant mobilizing role in the integration of the interior and exterior space. The best example of such spaces is in traditional Iranian architecture. Semi-open spaces have played an important role in traditional architectural spacing in the past and in the Iranian traditional architecture. The appearance and geometry of these spaces get their models from the features of their own surroundings, and these areas have hosted a range of individual and social activities. This character has been constructed simultaneously in the building in past architecture; modern semi-open spaces do not invite such human interactions and past times as they did in the past and generally lack architectural value. Today, semi-open spaces can be a good way to link human life with nature [1].

In Iranian homes, there is always a space between open and closed spaces. Indoor spaces between closed and open spaces are spaces that provide a variety of facilities for activities of the home and the community; such a presence offers a new connection with nature, light, and climate. Indoor spaces in Iranian homes have varied sizes and functions and are as important as open and closed spaces [27, 28]. Having said that, the presence of indoor spaces in the space organization of contemporary houses has dropped significantly. In some cases, the presence of a porch toward the yard creates a usable connector space between the open and closed space. In the upper floors, the covered terraces and balconies are the covered spaces. These spaces usually are not used completely, and in many cases they are presented as an abandoned floor. Although the balconies and terraces today are located along closed spaces, people prefer not to occupy them due to an undesirable view. Rather, residents typically use balconies as a storage. Nevertheless, these balconies can provide both the perspective and light from the northern and southern fronts for closed areas of contemporary houses [17] (**Figure 7**).

2.3 Designing private open space (balcony)

Many resources that provide designers with information on how to design spaces also include a series of criteria and components about the design of balconies; for example, Neufert et al. stated that balconies are effective factors in increasing the attractiveness of residential units. Balconies also create a space for some activities, such as outdoor play area for kids that can be easily monitored [29]. Furthermore,

Figure 7.
Areas that are located between the inner and the outer spaces of Iranian houses are highlighted in the images, adopted from [9], analyzed by authors.

balconies can be used to rest, sunbathe, study, eat, etc. In addition to the bio-functional aspects required, a space for the flower box in each part of the balcony is needed [30].

According to the Housing Design Handbook on Balconies Under Regulation and Principles, balconies should cater to the following uses: children's play space, a space for drying clothes, a space for sitting, gardening and planting, keeping pets, etc. [31].

Time Savers Standards for Building Types presents residents' opinions on what a balcony is ideally used for: people who have discussed much about the balconies and their positive role have emphasized the pleasure of using and sitting outside. In addition, they emphasize the visual expansion of life space and the opportunity to grow plants and use the balconies to store equipment [32]."

Therefore, the following uses are extracted from the studies on the design of balconies: "children playing, relaxing, sunbathing, studying, eating, keeping flowers and plants, drying clothes, sitting, keeping animals, and warehousing equipment." However, the factors mentioned are very general and it cannot be said that it is definitely welcomed by the residents in the specified neighborhood in which the research is carried out. Therefore, it is necessary to obtain an estimation of these factors and the factors extracted from the interview section, evaluate them quantitatively in the statistical population, and assess their external validity.

2.4 Description of theoretical model

2.4.1 Influential components in designing private open areas in residential buildings

The collection method of influential factors in the design of private open areas was discussed in the previous section (interviewing the residents of Kuye Nasr residential buildings and surveying balcony design studies). All collected factors are categorized into six groups as dependent variables, which can affect the quality of a space like a balcony. The components are classified based on a theoretical framework shown in **Figure 8**.

2.4.1.1 Esthetic aspects

This is a system based on balance and harmony. Regularity, balance, and fit are perceived and viewed automatically by the viewer and recognized for its beauty [19]. The beauty factors of balcony and the presence of greenery on the balcony belong to this category.

2.4.1.2 Behavioral aspects

These aspects include the hierarchical ordering and functional domains, access hierarchy, spatial hierarchy, and the hierarchy of various functions forming the communication between inside and outside spaces [33]. The functional structure of space affects the occurrence of behaviors in the environment and is simultaneously affected by the behaviors and anticipation of their occurrence [11]. Hanging clothes, sitting in the open air, cooking or barbecuing, smoking, using tools, entertaining children, enjoying a safe space, and escaping people are in this category.

2.4.1.3 Environmental aspects

The environment and behavior are so intertwined that it is difficult to separate them. Therefore, human behavior should be defined in an environmental context. In the past, the environment was considered as an independent variable that

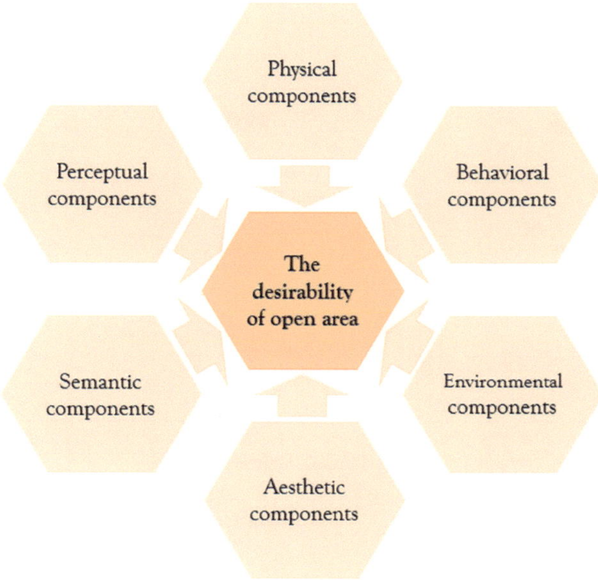

Figure 8.
Influential components in increasing the desirability of private open spaces.

affected behavior, shaped it, and created it. Consequently, one of the implications of this traditional approach is that human authority over the environment must be limited and the environment must be coordinated with people in a constant and unchanging form. Recently, however, research has been emphasized to have flexible and variable designs for the environments. In fact, people are now the cause of environmental change and are not limited to environmental influences [34]. Climate factors, tranquility and comfort of the balcony, and balcony orientation fall into this category.

2.4.1.4 Physical aspects

Physical factors include physical and visual order, balance in open and closed environments, walls, and physical bordering structure [35]. Proportional factors and the area of the balcony are included in this category.

2.4.2 Perceptual-semantic aspects

The most common approach to human psychology is one that considers several inner processes: perceptual reactions to the environment (how people understand and organize environmental stimuli and react to them), emotional and motivational states associated with environmental stimuli (psychological pressure and negative or positive emotions), and cognitive reactions to the environment (estimates of affluence, complexity, and meanings of the environment). This component considers humans as beings that have internal processors and deals more with mental and psychological processes than with obvious behavioral responses [34]. This factor is related to the human mental perception of the living environment. Visual communication, scale, and proportions, creating the sense place attachment and the sense of separation from space, visibility, and perspective from outside to inside and from inside to outside belong to this category of communication. The perception of sky, earth, buildings, surrounding spaces and, in general, the surroundings and relationship with them actually develop through giving meaning

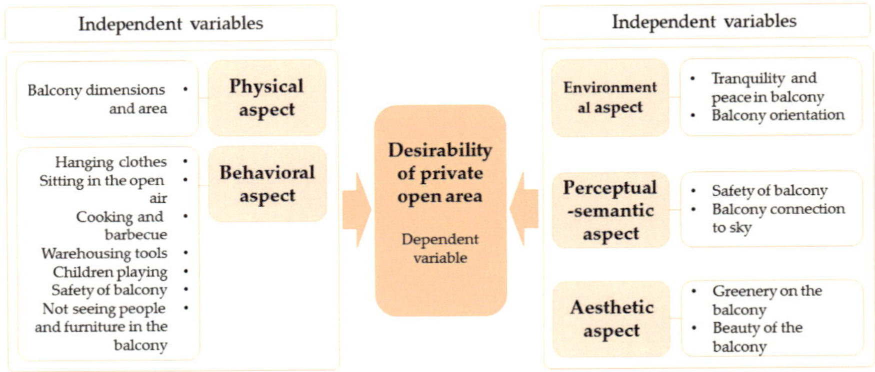

Figure 9.
Theoretical model of desirability of private open areas including five main aspects including physical, behavioral, environmental, perceptual semantic, and esthetic aspects.

Statistical methods and tests	Application	Type of statistical method
Frequency and rate of frequency	Describing demographic variables, describing main questions and variables	Descriptive statistics
Spearman correlation, chi-square test, one sample chi-square, Friedman rankings	Testing hypotheses	Inferential statistics

Table 2.
Statistical methods.

to the environment. This component is completely subjective before becoming an objective one [11]. Factors such as the security of the balcony space and the connection of the balcony to the sky fall into this category.

Therefore, based on the title of this article, "Effective factors on desirability of private open spaces in residential buildings"—independent components in a six-class categorization, in line with the following model in **Figure 9**— have a linear effect on the desirability of private open spaces as a dependent variable.

The next section investigates the role of the abovementioned components as influential independent factors in the design of a desirable balcony and introduces the methods and statistical patterns used to explore it, a summary of which is shown in **Table 2**.

2.5 Case study details

Kuye Nasr is the name of one of the northwestern neighborhoods of Tehran located in district 2. The main street of Kuye Nasr (Gisha), stretching approximately 1.5 km, directly connects two main highways to each other. Kuye Nasr has well-organized even and odd streets with English architecture and a complete grid-system urban structure with a total of 41 streets.

The reason for choosing this neighborhood as the case study of the research is the systematic grid texture of the district, which caused the lands in this area to be segmented north-south (regular northern and southern streets). Hence, the research and presentation of this proposal is more targeted and systematic compared to situations where buildings are studied in organic texture or other urban structures. Therefore, by distributing the questionnaire in a homogeneous statistical society, the final result will be generalizable and will have a greater external validity in the statistical population (**Figure 10**).

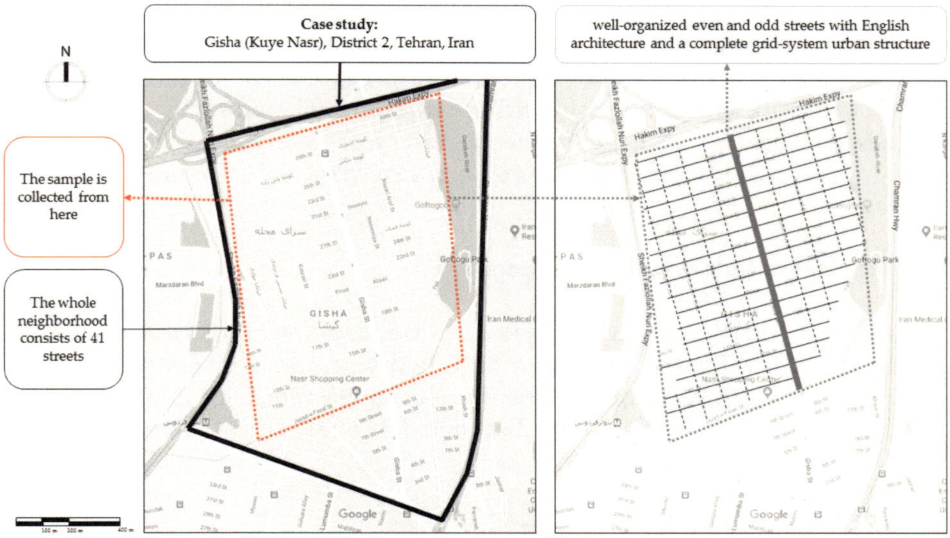

Figure 10.
Aerial map of the neighborhood, adopted from [9], detailed by authors.

3. Methodology

3.1 Type of research

The present research is categorized as applied research using quantitative method [36] and, using the results of the study, seeks to help improve and optimize the behaviors, products, structures, and patterns used by human societies.

3.2 Research method

The researcher photographed residential buildings in the intended texture to primarily extract the design factors of buildings. Next, using the interview method, the researcher collected the views and opinions of the residents of the residential buildings (the statistical population) regarding balconies; the optimal design factors for such spaces were collected by default. In order to ensure and assess the validity of the default factors, statistical studies (quantitative research method through the questionnaire tool and analysis of these data through SPSS 23) were implemented.

3.3 Research tools

This quantitative research method was adopted, in the form of a questionnaire distributed among people over the age of 20 who were living in the area. After the required data was collected, the principles, methods, and results of statistical analysis were used to study the validity and generalizability of hypotheses in the neighborhood.

3.4 Statistical population

The statistical population of the study includes the building residents in the Kuye Nasr neighborhood. The reason for choosing Kuye Nasr as the statistical population is the systematic grid texture of the neighborhood, which has led the lands to be segregated as northern and southern (regular northern and southern streets); therefore, the research and presentation of this proposal are more targeted and systematic compared to situations where buildings are studied in organic textures or other urban structures. Therefore, by distributing the questionnaire in a more homogeneous statistical society, the final result will be more generalizable and will have a greater external validity in the statistical population.

3.5 Statistical sample and sampling method

The research sampling method was a simple random sampling method. That is, from 20 alleys with even numbers and 20 alleys with odd numbers, two or three residents were selected randomly. After introducing the researcher and purpose of the study, the questionnaires were distributed. In total, 100 questionnaires were completed and analyzed.

3.6 Validity and reliability of the questionnaire

3.6.1 Validity

The validity of the questionnaire was tested by a non-statistical method; specifically, through confirmation by a number of experts. Experienced professors and PhD students in the field of architecture reviewed and approved each item on the questionnaire.

3.6.2 Reliability

Cronbach's alpha was used for the three-item questions. Cronbach's alpha coefficient was used to assess the reliability of the questionnaire. The most commonly used method for calculating reliability is the Cronbach alpha, which is called the alpha coefficient. The general rule is that the Cronbach alpha value of a scale should be at least 0.7 [37]. The Cronbach's alpha coefficient of the questionnaire is indicated in **Table 3**.

As can be seen, Cronbach's alpha value of the questionnaire is higher than the criterion 0.7, which confirms the reliability of the questionnaire.

Variables	Cronbach's alpha
Reliability	0.78

Table 3.
Reliability through internal consistency method (Cronbach's alpha).

4. Analysis and results

4.1 Analysis of the results obtained from field observations

A field study of the current balconies in the apartments or houses available in the site (**Figures 11, 12**) indicates that a large percentage of balconies on the first floors, which have easier access to the street or alley, are changed due to security concerns. That is, the balconies have lost their original form, creating a special

Figure 11.
(a) Northern apartment. (b) Southern apartment.

Figure 12.
(a) Southern apartment. (b) Northern apartment.

irregularity in the facade (as indicated in the figures below). The residents, for example, have fenced the balconies or have covered the open sides, so that they are safer and cannot be seen when they are on the balcony. Therefore, it can be concluded that it is best to design the balconies of the first floors in such a way that security is considered as a significant point in the design. It should be designed such that it does not require future changes by residents.

4.2 Analyzing the results of interviews

During their interviews, building residents were asked to identify their primary concerns regarding the balconies so that the researcher can start designing the questionnaire with an appropriate initial perception.

4.2.1 A list of primary requirements of residents regarding balconies is given below

The following requirements of residents in terms of using balconies are as follow: hanging clothes, planting flowers, cooking, having a spare section for putting kitchen appliances that are better kept in the open air, children's playing, having a view to beautiful landscapes, ventilation, and connecting balconies with kitchen (**Figure 13**).

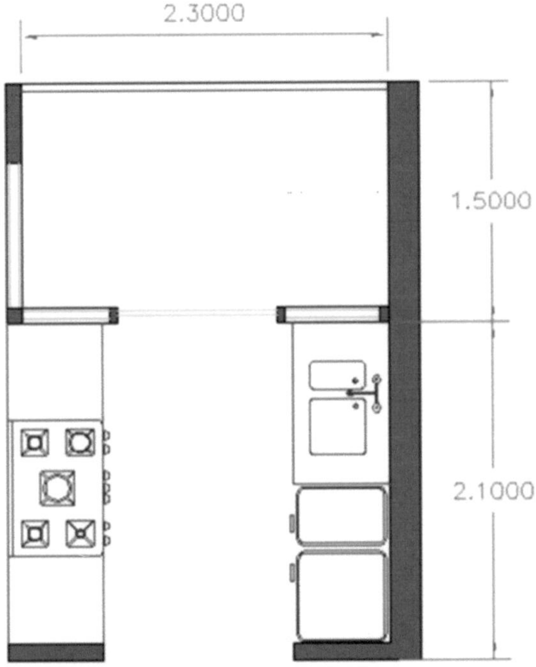

Figure 13.
Sample plan of a balcony connected to the kitchen, drawn by authors.

4.2.2 Connection between balcony and kitchen

The connection between kitchen and balcony is one of the main suggestions and requests made by Nuye Kasr building residents. During the interview process, the following reasons were extracted:

- The connection of the open space (balconies are almost the only open space available to the units) and cooking allows easier access to foodstuff and dining when barbecuing, as well as better ventilation of the kitchen while cooking.

- A space for sitting and drinking tea, having food, etc.; such a place is better to be connected to the kitchen.

- Access to kitchen appliances that are best protected in the open air.

- Proximity and easy access of the kitchen to the balcony to irrigate the plants on the balcony.

- A place for hanging clothes in the open air and easy access to the kitchen; nowadays, washing machines are embedded in this space.

All of the abovementioned arguments indicate that balconies are better placed and more efficient when connected to the kitchen. Thus, it can be concluded that in the intended neighborhood, the functional dimension of the balcony is highly regarded.

4.3 Analysis of the results

- Analysis of questions related to the main research hypothesis indicates that the presence of spaces, such as balconies, is essential in modern residential housing. A total of 99 out of 100 people who answered to the questionnaire agreed with the necessity of a balcony presence in their residential units. The results in **Table 4** also illustrate that the significance level of chi-square is less than the assumed value of 0.1, thereby confirming the hypothesis.

- Analysis of the demographic part of the questionnaire shows that tenants and homeowners do not differ in their need for a balcony, indicating the importance of designing balcony for residential houses of both types of economic situations. As shown in **Table 5**, the significance level of the chi-square independent test is 0.507 and is more than the assumed value of 0.10. This means that there is no significant relationship between the status of the residents of being tenant or owner and the need for a balcony.

- The activities performed on balconies nowadays are ranked in **Table 6**.

The comparison of the average ratings (**Table 6**) shows that the highest average rating for flowering and planting activity is at 5.55, which means that this is the most common activity on balconies. The activity of hanging clothes with an average of 5.43 and sitting in the open air with an average rating of 5.27 are the next highest ranking activities. The lowest average rating (3.09) is for children's activity, which indicates that children playing on the balcony have the lowest activity.

Type of relationship	Chi-square value	Degree of freedom	Significance level
Necessity of balcony	87.36	1	<0.001

Table 4.
Chi-square test: investigating the necessity of balcony presence.

Type of relationship	Chi-square value	Degree of freedom	Significance level
Between owner or tenant and necessity of balcony	1.356	2	0.507

Table 5.
Chi-square test: investigating the necessity of balcony from owner or tenant point of view.

Rank	Dimensions	Average score
1	Plants	5.5
2	Hanging clothes	5.43
3	Sitting in the open air	5.27
4	Enjoying view	5.12
5	Cooking and barbecue	4.80
6	Smoking	3.49
7	Storage	3.25
8	Children's activities	3.09

Table 6.
Friedman test: ranking of activities carried out on the balcony.

5. Satisfaction measures for designing private open spaces and future studies

It should be noted that considerations such as balcony dimension and area, the presence of greenery on the balcony, the connection of the balcony with the sky, comfort and tranquility of the balcony area, balcony space security, not being observable on the balcony, functionality of the balcony, and the beauty of balcony are the influential factors in the satisfaction of residents of buildings of their desirable private open area.

This topic of desirability and sustainability of private open areas can be extended further based on the following directions:

- Implementing new digital technologies including the Internet of things (IoT) [38] and geographic information system (GIS) [39–41] for monitoring and evaluating the space quality;

- Adoption of open space desirability measures in different types of buildings in dense urban areas [42];

- Challenges and opportunities for sustainable development of private open spaces to sustainable development of societies [43, 44];

- Investigating the process, factors, barriers, drivers, and decision makers using NVivo for analyzing the desirability of open spaces [45, 46];

- Analyzing the trend of private open spaces toward sustainable smart cities [46];

- Promoting more integrative strategies between stakeholders to improve the quality of private open spaces [47, 48].

6. Conclusion

This study tries to find ways to optimize the private open spaces of residential buildings; to this end, the required results were obtained through the analysis of the data and with the help of quantitative methods. The results are presented in the form of logical propositions. Firstly, this study recommends that the designers of the Kuye Nasr residential buildings consider security issues so that first-floor residents can use such spaces without the need for changes or extensions to the building's facade. Secondly, the functional features of balconies are of great importance to residents, and thus, its desirability factors should be paid close attention. Thirdly, the kitchen is the best and most welcoming space to be connected to the balcony. The kitchen offers the most suitable space connection in the buildings, in line with the functional aspect of the balcony. Fourthly, neighborhood building residents advocate for desirable factors including the presence of greenery on the balcony, the connection of the balcony with the sky, the tranquility and comfort of the balcony space, the security of the balcony, not being observable on the balcony, the functionality of the balcony, and beauty of the balcony. Finally, the present study ends with this question for future researchers: to what extent will the desirability of different parts of Iranian homes, which have been forgotten today or lack suitable quality, help improve the quality of their living spaces.

Author details

Reyhaneh Karimi[1*], Behnaz Avazpour[2] and Samad M.E. Sepasgozar[3]

1 University of Science and Culture, Tehran, Iran

2 University of New South Wales, Sydney, Australia

3 Construction Management and Property, Construction Project Management, University of New South Wales, Sydney, Australia

*Address all correspondence to: reyhanehkarimi1992@gmail.com

References

[1] Aliniay Motlagh, A., The connection between inner and outer space in new residential buildings, in faculty of fine arts. Tehran Univesrity. 2013

[2] Milanović D, Vasilevska L. Influence of private open spaces on the quality of living in low-rise high density housing. Facta Universitatis, Series: Architecture and Civil Engineering. 2018;**19**:293-305

[3] Mirmoghtadaee M. Process of housing transformation in Iran. Journal of Construction in Developing Countries. 2009;**14**(1):69-80

[4] Chau C-K, Tse M, Chung K. A choice experiment to estimate the effect of green experience on preferences and willingness-to-pay for green building attributes. Building and Environment. 2010;**45**(11):2553-2561

[5] Memarian G. House Typology in Iran (with Special Reference to Shiraz). Manchester: The University of Manchester; 1998

[6] Luttik J. The value of trees, water and open space as reflected by house prices in the Netherlands. Landscape and Urban Planning. 2000;**48**(3-4):161-167

[7] Bolitzer B, Netusil NR. The impact of open spaces on property values in Portland, Oregon. Journal of Environmental Management. 2000;**59**(3):185-193

[8] Soflaei F, Shokouhian M, Soflaei A. Traditional courtyard houses as a model for sustainable design: A case study on BWhs mesoclimate of Iran. Frontiers of Architectural Research. 2017;**6**(3):329-345

[9] Space Separators. 2019. Available from: www.pinterest.com

[10] Einifar A, Ghazizadeh N. Typology of residential complexes of Tehran with open space measure. Arman Shahr. 2010;**5**:35-45

[11] Einifar A, motlagh AA. Explaining the concept of outside-inside in In-between spaces of apartment housing “the case study of balcony in three types of Tehran residential complexes”. Journal of Fine Arts—Architecture and Urban Development. 2014;**19**:55-66

[12] Badeie N. Walls-privacy connection. In: Faculty of Fine Arts. Tehran: Tehran University; 2002

[13] Rezakhani Z. An introduction to the concept of joint in architecture based on heidegger methodology of word etymology. Journal of Iranian Architecture Studies. 2013;**5**:101-114

[14] Mir Shahzadeh S, Eslami G, Einifar A. The role of boundary space in the creation of meaning. City Identity. 2013;**9**:5-16

[15] Alexander C. A Pattern Language: Towns, Buildings, Construction. California: Oxford university press; 1977

[16] Bourdieu P. Distinction: A Social Critique of the Judgement of Taste. USA: Routledge; 2013

[17] Haeri Mazandarani, M., Home, Culture and Nature. 2009. Center of Studies and Research on Urbanism and Architecture of Iran

[18] Gehl J. Life between Buildings: Using Public Space. Washington, United States: Island Press; 2011

[19] Grütter JK. Grundlagen der Architektur-Wahrnehmung. Stuttgart: Springer; 2014

[20] Von Meiss P. Elements of Architecture: From Form to Place. UK: Routledge; 2013

[21] Arnheim R. The Dynamics of Architectural Form. USA: Univ of California Press; 1977

[22] Swapan A, Marinova D, Bay J. Understanding the importance of front yard accessibility for community building: A case study of Subiaco, Western Australia. Urban Science. 2018;**2**(2):41

[23] Reuben A et al. Residential neighborhood greenery and children's cognitive development. Social Science & Medicine. 2019;**230**:271-279

[24] Azad SP, Morinaga R, Kobayashi H. Effect of housing layout and open space morphology on residential environments–applying new density indices for evaluation of residential areas case study: Tehran, Iran. Journal of Asian Architecture and Building Engineering. 2018;**17**(1):79-86

[25] Bentley I. Responsive Environments: A Manual for Designers. USA: Routledge; 1985

[26] Lawson B. Language of Space. Berlin: Routledge; 2007

[27] Roodgar M et al. Sustainability, architectural topology and green building evaluations of Kashan-Iran as a hot-arid region. Procedia Engineering. 2011;**21**:811-819

[28] Memarian G, Brown FE. Climate, culture, and religion: Aspects of the traditional courtyard house in Iran. Journal of Architectural and Planning Research. 2003;**20**:181-198

[29] Kennedy R, Buys L, Miller E. Residents' experiences of privacy and comfort in multi-storey apartment dwellings in subtropical Brisbane. Sustainability. 2015;7(6):7741-7761

[30] Neufert E, Jones V, Thackara J. Architects' Data. England: Granada; 1980

[31] Levitt D. The Housing Design Handbook: A Guide to Good Practice. UK: Routledge; 2012

[32] De Chiara J. Time-Saver Standards for Building Types. USA: McGraw-Hill Professional Publishing; 2001

[33] Drucker-Brown S. House and hierarchy: Politics and domestic space in northern Ghana. Journal of the Royal Anthropological Institute. 2001;7(4):669-685

[34] Altman I. The Environment and Social Behavior: Privacy, Personal Space, Territory, and Crowding. USA, California: Monterey; 1975

[35] Einifar A, Aghalatifi A. The concept of territory in residential complexes: A comparative study of two residential complexes located in different height in Tehran. Journal of Fine Arts. 2011;**47**:17-28

[36] Sepasgozar SM et al. Implementing citizen centric technology in developing smart cities: A model for predicting the acceptance of urban technologies. Technological Forecasting and Social Change. 2019;**142**:105-116

[37] Brace N, Snelgar R, Kemp R. SPSS for Psychologists. London: Macmillan International Higher Education; 2012

[38] Shirowzhan S, et al. An integrated GIS and Wi-Fi based Locating system for improving construction labor communications. In: ISARC. Proceedings of the International Symposium on Automation and Robotics in Construction. Vilnius Gediminas Technical University, Department of Construction Economics; 2017

[39] Shirowzhan S, Sepasgozar SM. Spatial analysis using temporal point clouds in advanced GIS: Methods for ground elevation extraction in slant areas and building classifications. ISPRS International Journal of Geo- Information. 2019;**8**(3):120

[40] Shirowzhan S, Lim S, Trinder J. Enhanced autocorrelation-based algorithms for filtering airborne lidar data over urban areas. Journal of Surveying Engineering. 2016;**142**(2):04015008

[41] Shirowzhan S, Trinder J. Building classification from lidar data for spatio-temporal assessment of 3D urban developments. Procedia Engineering. 2017;**180**:1453-1461

[42] Thomas L, Cousins W. The compact city: A successful, desirable and achievable urban form. In: The Compact City: A Sustainable Urban Form. London, New York: E & FN Spon; 1996. pp. 53-65

[43] Costanza R et al. Building a sustainable and desirable economy-in-society-in-nature. In: Green Economy Reader. USA: Springer; 2017. pp. 367-454

[44] Madanipour A. Public and Private Spaces of the City. UK: Routledge; 2003

[45] Sepasgozar S, Davis S. Construction technology adoption cube: An investigation on process, factors, barriers, drivers and decision makers using NVivo and AHP analysis. Buildings. 2018;**8**(6):74

[46] Ullah F, Thaheem MJ, Sepasgozar SM. Sustainable smart cities: Evaluation of Australian practice. In: CONVR 2016 Proceedings of the 16th International Conference on Construction Applications of Virtual Reality 11-13 December 2016, Hong Kong. 2016

[47] Grose MJ. Changing relationships in public open space and private open space in suburbs in South-Western Australia. Landscape and Urban Planning. 2009;**92**(1):53-63

[48] Dair CM, Williams K. Sustainable land reuse: The influence of different stakeholders in achieving sustainable brownfield developments in England. Environment and Planning A. 2006;**38**(7):1345-1366

Section 2

Technology Applications in Construction

Chapter 7

Automating the Chaos: Intelligent Construction Contracts

Alan McNamara

Abstract

At the centre of all construction projects is the contract between the client and contractor and, as any construction project is a relatively complex process, the industry has demanded contracts of greater sophistication as the sector has evolved. The construction industry has a reputation for being adversarial and motivating dispute and the deep-rooted cultural aversion to trust. It is postulated that the solution to the trust issue is to make contracts trust-less. Truly autonomous, intelligent contracts would minimise the need for conventional human management. Put simply, intelligent contracts (or 'iContracts') are computer protocols that facilitate, verify, or enforce the negotiation or performance of a contract, or that obviate the need for a contractual clause. The proposal of embedding the terms and conditions of an agreement into a digital entity contrasts immensely from a traditional paper contract which is generally only used only as a reference when parties are in dispute. By creating an all-encompassing contract process—that: ensures all parties adhere to the terms agreed; offers protection of payment, insurance and data; as well as the potential to increase efficiency and reduce risk—it should make the successful implementation of iContracts the top priority for the construction industry.

Keywords: iContracts, intelligent contracts, digitalisation, automation, blockchain

1. Introduction

Globally, construction spending is projected to reach US$12.4 trillion by 2022 [1]. In the United Kingdom alone, it is estimated £600bn will be spent over the next 10 years on public and private infrastructure resulting in efficiency and productivity improvements in the delivery of construction projects becoming strategic priorities for the UK Government [2]. The construction sector faces many challenges including; low productivity, poor regulation and compliance, lack of trust, inadequate collaboration, information sharing, and poor payment practices [3, 4]. Whilst productivity is a major problem it also presents as one of the biggest areas for potential improvement with McKinsey Global Institute [5] reporting a global productivity gap of $1.6tr that can be addressed by improving the performance of the industry. The industry is perceived as slow to innovate, particularly in its adoption of digital technology but digital transformation is slowly gaining traction with increased use of Building Information Modelling (BIM) and emerging technologies based on Blockchain, Internet of Things (IoT) and Smart Contracts being touted as a solution to the industry's problems [6]. However, the integration of these digital concepts and technologies has not yet been achieved due to the embryonic nature of the field with further developments required to build a case for widespread adoption.

The construction sector is becoming more digitalised with BIM being the main catalyst for digital transformation seen in the sector in the last 15 years [7]. Smart contracts are seen to be one of the key complementary concepts to BIM due to the increased capabilities the automation of contract clauses will afford any user [8]. New digital technologies are emerging to address some the key concerns hindering collaboration in the construction industry and have the potential to change the way the sector operates, leading to better auditability and traceability encouraging more collaboration and information sharing [9]. In an industry that has historically lacked technological advancement and innovation, the construction industry is slowly being dragged into the digital age but there are many challenges to be addressed before true digital transformation is realised. Examining barriers to implementation and prevailing stakeholder attitudes are also crucial in envisioning this departure for construction and engineering contracts.

This chapter will define the iContract technology along with a state of the art of current research. It will highlight the challenges and barriers implementation of the concept will face, along with the opportunities it would bring as the construction industry enters further into the digital age.

2. Background

2.1 Striving for collaboration through contracts—is standardised really the standard?

The level of collaboration and the nature of a contractual relationship are usually dependent upon the proximity of the parties. Historically only simple contracts are adopted where person to person relationships are created and where high levels of trust already exist. The evolution of society and commercialism has now meant that more personable relationships are harder to come by or do not exist at all, and it is necessary to express more aspects of any commercial relationship through more defined contracts. Open competition and globalisation of contracting activity can be a barrier to developing trust leading to the creation of more sophisticated and detailed contractual arrangements in lieu of existing trusting relationships.

Standard forms of contract, so lauded by the construction industry, have the downside of restricting the freedom to contract upon one's own terms and therefore are contrary to an open market approach. Standard contracts certainly have their place in bringing consistency to the industry's contractual practices, but it must be remembered that most construction projects are a one-off prototype that require more specific terms. Standardised contracts would have a greater impact in a more heavily planned and regulated sector, which the construction industry would certainly benefit from but does not possess. The alternative, to adopt non-standard forms of contract, produces higher transaction costs and loses the benefits of familiarisation leading to extended contract formation and negotiation periods. However, digitalisation offers solutions to these problems and can improve the efficiency and communication process promoting instant collaboration and mitigating the level of disputes.

The reliance upon looseness of language and implied terms in an attempt to achieve flexibility, and reduce the content of contracts, is a practice that offers the double-edged sword of alternative interpretation and opportunity to act against the spirit of collaboration. A road well worn by construction companies put under pressure. This practice is only worth considering if the implied terms are such that they are well known and fully understood by all parties, which is generally only built from past engagements. Digitally coding such terms may assist but there are limits to the help this would offer and the logical coding of flexible arrangements has its challenges.

There is an apparent trade-off between a standard contract which in a general sense is understood but does not meet the real needs of the parties, and one that meets this need by producing a bespoke and detailed contract. Some commentators have argued that users do not 'really' understand standard forms of contract. The compromised situation of producing bespoke contracts with common clauses may actually improve standardisation and understanding when compared to the current practice of bastardising a standard form of contract beyond recognition to suit one's needs.

2.2 Industry 4.0—the digital dawn

Industry 4.0 is the era where computer power becomes more embedded within society and possibly even inside human beings [10] and has been used to describe the 4th Industrial Revolution by pointing out its huge technological potential, comparable to technical innovations which led to the first industrial revolutions [11]:

1. the field of mechanisation;

2. the use of electricity and;

3. the beginning of digitisation.

In the United States, construction is the least digitalised sector and has been publicly perceived as an industry with poor productivity and a low level of technology implementation whilst the financial industry and business services show the highest levels of digitalisation [12]. Construction is one of the largest and most significant industries in Australia, contributing to economic growth and jobs (ca. 9% of jobs). In 2010, the industry was considered the fourth largest industry in the country [13].

The Farmer Report [3] outlines a case for digital disruption in the UK construction industry and sees the following as 'critical symptoms of failure and poor performance':

- Low productivity
- Low predictability
- Structural fragmentation
- Leadership fragmentation
- Low margins
- Adversarial pricing models and financial fragility
- A dysfunctional training funding and delivery model
- Workforce size and demographics
- Lack of collaboration and improvement culture
- Lack of R&D and investment in innovation
- Poor industry image

This is set against an industry which is not applying for billions of pounds of R&D Tax Credits, set up by the UK Government to stimulate innovation [14].

Joseph Schumpeter explained the challenges when new innovations disrupt the traditional way of life as, 'Creative Destruction... Just as the day Samuel Morse invented the telegraph was a bad day for the horse back messenger, significant technological disruptions can be destructive for older more established workflows that do not adapt' [15].

From a technical point of view, Industry 4.0 can be described as the increasing digitisation and automation of the manufacturing environment as well as the creation of a digital value chain to enable the communication between products and their environment and business partners [16]. The industry specific definition of the Industry 4.0 concept for construction comprises a large range of technologies to enable the digitisation, automation and integration of the construction process at all stages of the construction value chain. Central technologies like BIM, Cloud Computing or the IoT are only a few of them some of the main, well known technologies. Typical base technologies and concepts of Industry 4.0 are: The IoT/Internet of Services (IoS), Cloud Computing, Big Data, Smart Factory, 3D-Printing and the Cyber-Physical Systems (CPS) or Embedded systems. There are also emerging technologies: Augmented Reality (AR)/Virtual Reality (VR)/ Mixed Reality (MR) and the Human-Computer-Interaction (HCI) are major components of Industry 4.0 to enable a digitised and automated construction environment.

These technologies are at different levels of maturity. Scheduling, communications and BIM, among others, have reached market maturity and thus are currently available whilst others, such as IoT and AI, are still at the formative prototype stage. Despite the maturity and availability of many technologies, their widespread adoption by the construction sector has been slow but there are signs that there are practical ways for the successful adoption of new technologies to digitise and automate the construction process. Even though these technologies can have hard to predict cost savings due to the increasing need for data security and data infrastructure. From the technical point of view, there are several unsolved problems and challenges to be met due to the lack of standards for many new technologies and the higher standard of IT infrastructure required to run them. Regulatory compliance and legal uncertainty are other issues to be considered. Considering all these challenges, it is clear that companies must be motivated to commit to the adoption through government initiatives or funding programs [17].

The adoption of the Industry 4.0 concept would help the construction industry transform to a technology-driven sector and help keep up with other industries in terms of performance improvement. The Mckinsey Group 2016 report suggests that, 'the adaptation of currently demonstrated automation technologies could affect 50% of the world economy, or 1.2 billion employees and USD $14.6 trillion in wages. Just four countries—China, India, Japan, and the United States—account for just over half of these totals' [18]. The digital revolution is here, and the construction sector needs to be on board, or it will be left behind.

The increased rate of technology in society today is undeniable with the number of sensors in devices set to pass 25 billion mark by 2020 [19]. As the huge amounts of data that IoT will bring to the table will have to be managed, the implementation of Big Data solutions can help to collect the right data from all data-generating devices and to make them accessible to stakeholders [20]. The analysis of big data allows the identification of patterns and probabilities of construction risks for performance optimisation in real-time or on future projects [21].

The harnessing of all this data to a central source of analysis is where the iContract technology can excel. By having an endless supply of sophisticated data along with the ability to analyse at computational speed, an iContract could address the current problems with communication, accuracy of data and speed of action, rife within the construction sector due to the reliance on paper contracts administered manually.

3. The current digital construction landscape

3.1 Current and emerging technologies—new kids on the blockchain

The blockchain concept was brought to prominence in 2008 when Satoshi Nakamoto published the white paper 'Bitcoin: A Peer-to-Peer Electronic Cash System' which based the Bitcoin crypto currency concept on blockchain technology. A blockchain is a ledger, or a database of transactions recorded by a network of computers' [22]. Often referred to as distributed ledger technology, transactions are grouped in blocks and the chain forms the history of these transactions (the blockchain). It is widely believed to have been created as a way to distribute crypto-currency in a way that maintains publicly, and by multiple people a record of the transaction [23]. Trust is built into the technology through its decentralised nature and basis of consensus representing a paradigm shift from trust to a 'trust-less' system in which third parties become redundant. Blockchain-based Intelligent Contracts would therefore be more sophisticated, following the stigmergic ideology first coined by Pierre-Paul Grasse in his research on termites, and qualify as computer software code that is autonomous and independent as it cannot be controlled by any one entity [24] (**Figure 1**).

A blockchain is a distributed database (ledger) that maintains a list of records or transactions [1]. These records are called blocks and each block has a timestamp, a link to the previous block, and contains the history of every previous block that came before it. This 'chains' the blocks together, hence the name 'blockchain'. The whole system is fully encrypted with every transaction given a unique cryptographic signature (called a hash function) that is easy to verify and nearly impossible to falsify. This is because the blockchain is hosted by not one single source of truth, but rather

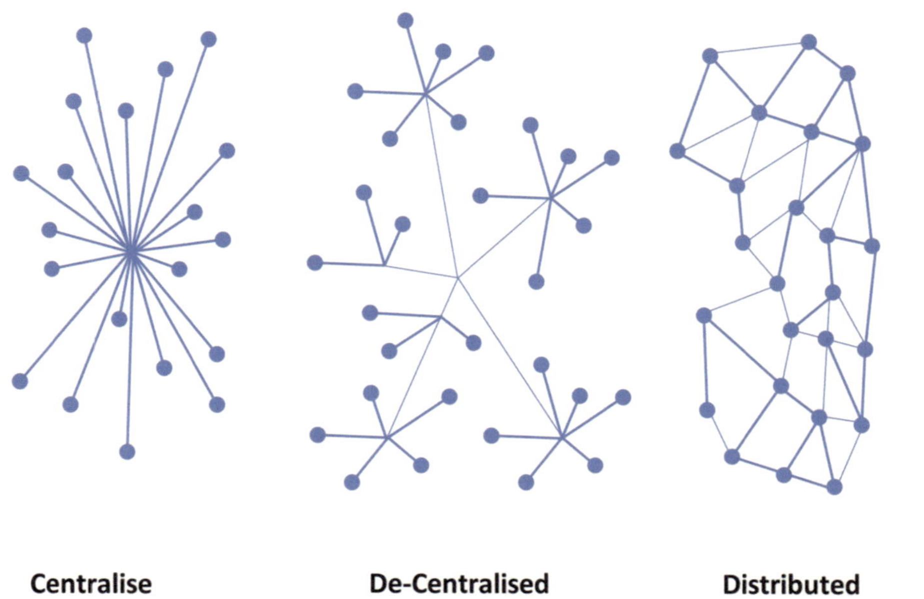

Figure 1.
Centralised network vs. a decentralised network vs. a distributed network (blockchain).

a network of computers. Each computer is a 'node' in the system and each node monitors every other node on a continuous basis verifying consistency of information. Each node checks its local record (block) with every other block in the chain and if it discovers a discrepancy, it looks for consensus and, if necessary, replaces the block with the consensus. This means that to alter a record it is necessary to simultaneously change the majority of the nodes in the system as oppose to one single source. It is not impossible but increases in difficulty as the length of the blockchain grows. The use of the blockchain has moved on from simply being the platform for crypto-currency, to ideas of cheaper transaction processing, crowdfunding and smart contracts. Blockchain 'holds promise for being the latest disruptive technology,' [22]. The heavily transactional nature of any construction project would lend itself to optimisation through a blockchain process allowing for 'automated' trust to be engrained into the acceptance of the next wave of digital technologies. **Figure 2a** shows a typical transactional/relationship map of a construction project and how the general contractor is often the central point to any operation. The disjointed nature of the relationships in this traditional process lends itself to inefficiency and error due to the linear transaction paths. **Figure 2** shows how a distributed network with an unrestricted flow of transactions verified by an immutable blockchain would remove reliance on any one party to validate or confirm data.

BIM is currently the expression of digital innovation within the construction sector. If BIM is the main enabler for promoting collaboration, information sharing and data management, blockchain is a possible solution to eliminating the trust element related to the vast transactions of data [1]. Although it is generally accepted that BIM would benefit from integration with blockchain technology, there is a consensus that the degree of collaboration enabled by Level 2 BIM is insufficient, and BIM usage must reach Level 3 before this can be realised [23]. Level 2 BIM is a collaborative way of working, in which 3D models are created by the various disciplines engaged on a project according to a set of guides, standards and specifications. Level 3 BIM relates to open process and data integration enabled by web services and also focuses on working within a new contractual framework that promotes consistency, clarity, openness and collaboration. With the expected increase of more sophisticated data becoming available through emerging IoT technologies the possibility to realise level 3 BIM may be soon.

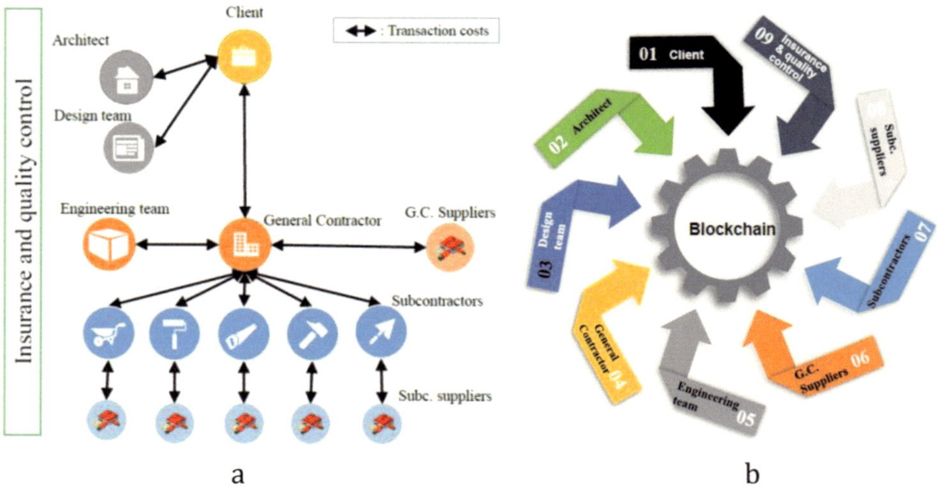

Figure 2.
(a) (The contractual relationships and the flow of money) and (b) (blockchain-based construction). Images from 'The Potential of Blockchain in Construction' [25].

According to a recent report by McKinsey & Company, IoT technology can offer savings of between $160 billion to $930 billion annually on construction sites and other related industries such as the resources sector [26]. It has been reported however, that uptake of the technology within the construction sector has been slow as within the research on IoT within construction, there are many articles to promote IoT without a tremendous amount of real world construction case studies [17].

IoT is described as a combination of sensors such as Radio-frequency identification (RFID), other communication devices, cloud applications and business intelligence technology. With billions of low-cost sensors becoming available, Big Data will flow from previously untapped sources to help improve decision making. The sensors can be embedded in physical objects, such as vehicles and heavy equipment or robotics and even building components themselves with the ability to connect to the Internet and communicate data [5]. These sensors can then send performance data to any other source to be analysed.

In the asset management phase of a project's lifecycle the technology would help the Asset Manager to reduce reactive maintenance practices in favour of predictive maintenance by allowing maintenance managers to repair the components and equipment before damage occurs from the information flowing automatically from sensors imbedded in building components, thereby reducing costs [17].

The obvious evolution to iContracts in maturing the level of automation is already well documented to be heavily linked with the development of blockchain technology and cryptocurrencies [8]. This weighs heavily on the Code is Law concept [27] that has become popular following the widespread deployment of the internet and society's dependence on digital technologies [28].

Smart contracts and blockchain are technologies that have been made a significant contribution in various industries such as the healthcare and finance sectors [29, 30]. The construction industry requires a disruptive influence in order to force it down the route of digitalisation and adoption of cutting edge of technologies seen in other sectors. The BIM platform has set the foundation for iContracts to launch from and provide a completely digital construction process solution, but the technology faces many barriers to achieve implementation. A popular, and valid, view is that construction contracts would be too complex and difficult to code and that they would not cope with legal wriggle-room words such as reasonable and satisfactory often relied on in the sector. This is where further investigation into a semi-automated approach, where the verification of tasks is still carried out by a human interface yet executed via the iContract, is needed to mitigate the discomfort of users handing over control to an untried technology.

As BIM and data management technology drive these new approaches in the construction industry, there will be a need not only to consider the contractual regime, but also to challenge the traditional competitive procurement and tendering processes with more collaborative structures and approaches [31]. By using Blockchain and BIM in tandem, along with other quickly advancing technologies, there is an opportunity to create leaner procurement methods which pushes the collaborative agenda. This will result in cost reductions by removing the multitude of intermediaries currently embroiled in the traditional process and will give a client greater control and transparency of cost, time and scope [32].

3.2 From paper to smart to intelligent contracts

Law and management were traditionally separate disciplines in the construction sector until realisation that law and management could co-exist which has driven the contractual developments in the last few decades championed in part by the

Latham and Egan agendas of the 1990s. The next breakthrough in the field is likely to involve the connection of law, management and technology.

Traditional contracts which are formed through a prolonged period of negotiation between parties, have elaborate drafting phases and require the contribution of lawyers in order to establish the contracting party's obligations. The manual nature of the traditional process is rife with inconsistent and ambiguous drafting practices that are the source of the majority of dispute in the industry. Poor contract administration continues to be the number one cause of construction dispute with errors and translation of contracts being within the top three causes [33]. Unfortunately, the onerous nature of contract administration leads to many human errors which have a detrimental impact on construction projects due to the inefficient nature of manual contracts.

Smart contracts are considered a key influential development that will support Britain's achievement to becoming a digital economy as set out in the government report—Digital Built Britain [34]. Smart contracts have the potential to remove the need for a trusted third party to administer a contract in a truly autonomous state by integrating Building Information Modelling and the IoT, to inform the smart contract of actual progress and performance. The concept of smart contracts, conceived in 1994 by Nick Szabo, is a computerised transaction protocol that executes the terms of a coded contract [35]. The general objectives of smart contract design are to satisfy common contractual conditions, minimise exceptions both malicious and accidental, and minimise the need for trusted intermediaries [40]. Related economic goals include lowering fraud loss, arbitration and enforcement costs, and other transaction costs. The effect of such contracts on contract law and economics, and their opportunities were said by their originator to be 'vast but little explored'. Szabo had a broader expectation for the smart contract concept that through specification of clear logic, and verification or enforcement through cryptographic protocols and other digital security mechanisms, smart contracts might offer an improvement over traditional contract law for efficiency in initiating contractual clauses that could be brought under the dominion of computer protocols.

Smart contracts have also been defined as 'Contracts that are fully executable without human intervention' [36], or 'Self-enforcing, monitoring external inputs from trusted sources in order to settle according to the contracts stipulations'. The key characteristics of smart contracts were described by the Norton-Rose-Fulbright report [37]:

- Digital form: it is in code form
- Embedded: contractual clauses (or equivalent functional outcomes) are embedded as code in hardware or software
- Performance mediated by technological means: the release of payments and other actions are enabled by technology and rules-based operations
- Irrevocable: once initiated, the outcomes for which a smart contract is encoded to perform cannot typically be stopped (unless an outcome depends on an unmet condition). It performs automatically.

Smart contracts translate the legal terms and processes into software code; therefore any contractual response is the outcome of the programmed code. Once initiated, it typically cannot be stopped or reversed once commenced without built in protocols allowing for alterations. Artificial Intelligence (AI) also has the opportunity to be included in smart contracts to assist with decision making as the technology develops [38]. There are several levels of smart contract models, ranging from a fully autonomous contract where the conditions are entirely in code, to a semi-automated natural language contract where only the payment mechanisms are encoded.

The application of smart contracts in the financial sector would appear easier to establish given the relative straight forward nature of the instruments involved as oppose to the construction sector where every construction project is different, with a specific design and scope of works, this type of contract drafting is complex, and trying to account for all contingencies is not possible. The general rule of thumb is that the longer the contract, the less straightforward its automation.

iContract is the term used when a contracts purpose is to manage itself [23]. An iContract will set out the requirements and decision inputs (hold points) in order to start a series of if/then that will execute the terms of the contract between the client and different members of the project team; main contractor, sub-contractors and any consultants or specialists involved. The iContract clauses are executed when the coded contractual conditions are met allowing digital transaction information such as performance criteria, physical existence of materials on site and works complete to verify a payment amount to be embedded and automatically transfer among the contracted parties once the agreed parameters are met [8].

The 'black and white' or '1 or 0' execution of an iContract is a huge obstacle to overcome in adopting the potential technology due to the complexities of the construction process requiring judgement and discretion which would normally be handled through subtlety and refinement in the language of traditional contracts. These challenges along with the potential benefits of the concept are explained further in the next section.

4. iContracts—an outlook

4.1 Benefits and applications

4.1.1 Optimised contract formulation and negotiation

As clients rush to proceed with the construction of projects, the industry's complex contract formation and protracted negotiation process remains very time consuming and expensive. The current practice requires a great deal of resources for the negotiation of contracts and seemingly self-defeating nature of amendments. The benefits of standardisation in terms of ensuring minimum quality standards and the potential to reduce disputes through un-ambiguous contract drafting is apparent but presently where standard forms are used, they are often modified to the point where they are no longer standard negating the advantage of their purpose. The potential to optimise the drafting of contracts in a logical format that will avoid individuals interpreting contract clauses differently due to the way they are written would be a major advantage of iContracts.

An evolution towards the automation of the contract formation and negotiation process could not only reduce the expenditure of resources, but it would also alleviate the ambiguous nature of current contract drafting as the iContract would be more logical in nature. The possibility for a digital database of clauses and terms to be automatically recommended by the technology, based on criteria set by the user, would offer the opportunity to greatly reduce the drafting and negotiation period.

4.1.2 Contract admin efficiency

Poor contract administration continues to be the number one cause of construction dispute with errors and translation of contracts running as a theme within the top 5 causes [33]. The construction sector has a reputation for being adversarial and lacking in collaborative and efficient practices. The largest contributing factors to this problem are the behavioural and adversarial attitudes to the administration of what are generally convoluted and inconsistent contracts whilst the heavily onerous nature of contract administration detracts construction managers from the actual delivery of a construction project [39]. A reduction in human error can be achieved through the automation of tasks, using IoT sensors, artificial intelligence and smart contracts. Certification and verification of tasks through blockchain protected interfaces feeding an iContract would provide increased quality assurance for construction projects. Many contract admin tasks could be automated changing how organisations operate which would benefit the industry by speeding up the payment of funds linked to, or possibly embedded via crypto-currency with, an iContract.

The potential for the redeployment of people and resources to other tasks due to the efficiencies an iContract solution could offer would offer any project a huge benefit. iContracts would increase the collaboration, transparency and improving accountability and project control as workflows supported with an automated iContract would allow the waiting time for 'sign-off' to be eliminated as *input* for completed tasks links through an automatic *forward* in the communications protocol. This would bring efficiency to the project schedule by removing disruptive contract bottlenecks allowing continuance of work [1].

Whilst workflow and process control has already emerged through communication and document control systems such as *Aconex* and *Teambinder*, the execution of those workflows are still managed manually based on what is interpreted from the construction contract. An iContract could easily link to these systems and extract the necessary data thus populating the relevant workflows automatically and accurately. Any resultant transactions or tasks would then flow through an iContract giving it a contractual checks and balance function to every transaction on a project.

An iContract solution would alleviate the onerous contract administrative tasks currently handled manually allowing greater speed and accuracy of the process whilst diverting the effort of management to project delivery.

4.1.3 Improved communication, collaboration and trust

The developments and trends studied in the last decade have centred on partnering arrangements and the promotion of more collaborative working relationships. The barrier preventing these initiatives from gaining traction has been the propensity of the sector towards the distrustful and adversarial approach inherent in the industry. The mere term 'business ethics' has been deemed an oxymoron [23].

Lack of trust and limited collaboration between parties has been one of the most frequent cited issues within the construction sector with digital technologies being heralded as the ultimate solution to finally address these downfalls. As digitalisation takes hold the need to become more transparent, improve communication, and increase collaboration and trust is promoted between parties through necessity of adopting digital technologies [40].

Through automation, processes are clearer and more transparent by their nature allowing the trust between the contracting parties to improve. The idea that collaboration is part and parcel of the automated process and that, far from being a casualty, is part of the DNA of a potential iContract is something that could finally demand what has been sought for decades. The construction industry may well have exhausted its ability to collaborate through traditional mechanisms due to the human based factor allowing for digitalisation to disrupt.

4.1.4 Supply chain efficiency

Proof of source of materials can be achieved through digital technology with the potential to provide better record keeping across the entire supply chain for a project through a traceable, immutable digital capture of actions and transactions. This would give the ability to immediately pinpoint where problems exist should they arise as people are held more accountable for their actions through increased transparency.

Tracking of goods and services throughout the supply chain offering near real-time data as well as live data about components in the construction to the BIM model providing updates on the 'as is' state of the building can be achieved by a complete digital ledger solution. Performance and reputation ratings implemented through a digital supply chain would also be possible promoting strategic partnerships as trust would be auditable and earned, much like the driver ratings system found in Uber. The digital integration of real time analysis along with efficiency of the process through the digital procurement process that an iContract would drive could achieve an optimisation of the supply chain in an industry that is seen to be heavily fragmented. The possibility of actually removing intermediaries from the construction project supply chain has also been proposed by some commentators [41] as the structure of the British construction industry is shown to be dominated by main contractors who are essentially intermediaries between the owner and the lower supply chain relying solely on cash flow for profit.

4.1.5 Real time scenario analysis

Quick response rates are a crucial element for any construction project due to the ad-hoc nature of construction activities. A construction contract provides a mechanism for setting the original timescale as well as the direction to vary the relevant parameters upon the occurrence said events. Unfortunately, applying these mechanisms are incredibly onerous, time consuming and offer opportunity for inaccuracy due to the manual nature. The processing power computers could not only accomplish this task but also apply added analysis through prediction and remodelling of a schedule as eventualities occur.

An iContract could run scenarios for any scenario, be it legislative changes in requirements impacting on construction methodology or materials, to any number of events that arise on a construction site daily. By having an intuitive and sophisticated digital contract engine, a user could run a scenario, either as a simulated possibility or based on a real event with the iContract informing the user what the contract consequence will be. The iContract can then, acting in either an advisory or automatic fashion, execute the appropriate contractual workflow to remedy the situation. As the flow of data from external sources such as BIM models, schedules and cost control software becomes more reliable, the iContract can begin to run autonomously and begin to incorporate artificial intelligence in contract clash detection and forecasting dispute avoidance as trends are recognised. Optimising change management through the speed and accuracy of a digital solution would be extremely benefitable to the decision-making process in any construction management team.

4.1.6 Performance analysis and forecasting

Imagine we wanted to record the temperature on site an extreme climate location, every 5 min whilst pouring concrete for quality control purposes. This would be very labour intensive, error prone and tedious. IoT technologies have made measuring the physical world and submitting measurement data much easier and safer. Much like the pivotal role iContracts could play in real-time contract scenario analysis, the opportunity to analyse contract performance from real world data sets would allow the verification of real time performance against that of the contract requirement. The capture and analysis of performance data could capture upward or downward trends in contract performance immediately, or even before, the fact allowing the project team to address the situation.

4.1.7 Increased traceability and accountability

Current practice in the construction industry demands a need for accountability and for liability to assign to those who err with blame being attributed to the culpable party. The notion of moving away from this blame culture has been recognised as a step towards a more collaborative environment with the potential to encourage this using a central 'project insurance'. Having a central project insurance policy in place of the multiple policies normally encountered on a construction project forces all parties to work together through problems encountered due to the shared policy held by all. The iContract concept would complement this prospect and contribute a concise 'central truth' to any claims made to the project insurance from the data the iContract would hold, all in an easily searchable, digital format.

Traceability of every contractual transaction would be far more accurate and easily found due to the digital nature of the iContract. Not only would data of any project be searchable, the prospect of missing or incomplete data due to manual input would be minimised due to the automated nature of the administration process. Through blockchain technology, an iContract could create an immutable record adding more transparency to every transaction of a construction project.

4.1.8 Stability of payment process

Payments in construction contracts have long represented one of the biggest challenges for the industry [42] and iContracts have the potential to offer a solution to this fundamental problem. The alleviation of this through a transparent and reliable payment process would be welcomed and is one of the major benefits of the iContract adoption. A contract with a self-executing function making automatic payments upon completion of defined obligations, thereby speeding up payments for contractors removes the human element of wrong-doing so prevalent in the sector [43]. Combined with cryptocurrencies, the potential for guaranteed payments increases significantly.

An iContract could be comprised of not one but multiple mini-contracts, all self-executing, transferring data as they execute whilst generating the relevant payment once relevant parameters such as performance achievement have been met. An iContract could act as a 'trustworthy contract administrator by introducing an error-free process'.

Recent advances in cryptocurrency, big data sensors and project bank accounts could lead, at the very least, to a semi-automation solution to the payment function with completion of the work being verified either by a human or external data technology.

Through a central iContract system payment could automatically flow, not only to the head contractor, but also to subcontractors, consultants and suppliers, where cashflow is critical to the survival of construction businesses. An iContract would dictate when, where and how a party is being paid and what for based on a more accurate digital process. The services rendered by the contractor must obviously be made clear and the coordination of BIM, program schedule and cost model may be needed to triangulate the parameters for which the money is released. The capability for the iContract to be central to automated payments is something that will be a huge benefit to the industry as the capability for 'pay for work' becomes an instant process resulting in 'inch-stone' payments as oppose to the traditional longer milestone payment terms that are extremely onerous to many players in the industry [7].

4.1.9 Reduced dispute

Adversarial behaviour is common place within the industry with contracting practices being the main cause of dispute costing the industry hundreds of billions of dollars every year globally, with the average dispute costing US$42.8 million globally with the average length of disputes taking 14 months to resolve [33]. The potential to significantly reduce dispute is a major part of the business case for the adoption of iContracts as they have the potential to radically reduce the scope, type and size of contract dispute. The immutable nature of an iContract system would present a contract environment where facts are harder to dispute due to the optimised and efficient data management leading to less disputes. The onerous nature of human administration in any construction contract dispute is something that could be potentially alleviated through automation of the process.

4.2 Challenges and barriers and possible solutions

4.2.1 Innovation adoption in the industry

The construction industry is seen to be typically slow at adopting new technologies and historically resistant to change. Some commentators offer the opinion the sector is not yet ready for the level of collaboration and information exchange required for a digital automated contract to be successful [39]. Some believe that due to the technological state of the industry being insufficient, implementation of blockchain and other digital solutions is likely to be costly [40]. Generally, digital technologies are presumed to increase productivity, but this is not always the case as, if it is not combined with efficient and streamlined processes or when organisations lack a collaborative environment, it can struggle to make an impact.

The key to evolving construction will be having a robust enough central management system that will allow digitalisation and automation to flourish. The construction contract between the stakeholders of any construction project is the central point that all data must flow in order to be analysed and actioned. The construction contract must therefore have a greater capability in operating in a digital world, where the abundance of data to be considered will only continue to increase. The industry must overcome the institutional inertia that is ingrained its culture to present the appropriate digital environment for the iContract to flourish.

4.2.2 Handing decision making to an automated process

When contemplating the prospect of an iContract process the question of whether a party has the right to challenge any decision made by the automated process and what protocols would be required to retain a user's right to question a decision without negating the purpose of the automation in the first place. Commentators on iContracts have discussed the need for the ability to regain control of any automated system in the event of an unpalatable outcome to be perhaps agreed by all parties. Acceptance of the iContract certainly leans towards a semi-automated version of the iContract where the majority of the groundwork is carried out by the software with a human supervisor. This may offer an element of comfort as the concept evolves towards a more ingrained autonomous role in the future.

An element of surrender to an automated system has been identified as a limitation to the industry [39]. Traditional construction contracts require judgement and discretion which is extremely different to code. The benefits of iContracts are diluted by the logical '1 or 0' process it must rely on. The alternative is that computers are a tool and can perform a good deal of the repeatable aspects of construction whilst allowing for human input on the more sophisticated tasks or act as a hold point for any critical decision. This is the semi-automated position advocated as likely to be the work around in the short to medium term.

A phased based approach, much like the BIM levels, appears to be the likely roadmap with a semi-automated process being developed using existing contractual procedures. Identification of the processes that would achieve the greatest cost/quality/time saving, whilst achieving confidence in the process by giving an element of human control, should make the concept more appetising for the industry.

4.2.3 Technological and data requirements

BIM's establishment in recent years has laid the foundation for iContracts to operate. The counter-argument that iContracts would not need to align so closely with the BIM agenda due to basing themselves as not one multi-party contract but a collective of possibly thousands of contracts is also something to be considered.

Given the embryonic stage of the iContract concept and the lifecycle of new technology in general, it is expected that many of the challenges highlighted will be solved as existing technologies evolve. Through cloud computing it is possible to access and combine data from various emerging construction software applications through data virtualisation and an Application Programming Interface (API) that allows data from one application be used by another.

The appropriate regulatory and technological infrastructure must exist in order for the iContract to thrive long-term. Facilitating its adoption and integration with other established technologies such as BIM, scheduling software, communication and document control software and other Project Management tools will be critical to the success of any iContract platform. **Figure 3** shows the possible relationships between the physical world, the construction information environment and an iContract based on a blockchain platform.

4.2.4 Perceived legal inflexibility of a digital contract

The adoption of an iContract faces many challenges from a legal perspective as the irrevocable nature of a coded document poses problems in terms of satisfying the contracted parties that the coding is operating within the same parameters of a traditional manual contract. The element of trust required in the system is something that is not required within the traditional model.

Manual construction contracts deal with uncertainty by containing wording allowing a flexible approach to be taken when situations arise. One of the main perceptions of automated contracts is that they will be incapable of dealing with the 'wriggle room' that exist in traditional contracts. A computer programme is made up of algorithms which are essentially 'if x = y then z' and the ability for iContracts to deal with change and uncertainty will be a major barrier preventing their adoption. The difficulty in replacing subjective 'loose' wording with computer code is a huge challenge in order to cover the multitude of variables encountered on any construction project but this is again where a semi-automated human-interaction hybrid model may ease sceptics of the technology.

The recurring theme in most commentary on the automated contract subject is that construction projects are unique, and the size and complexity of projects

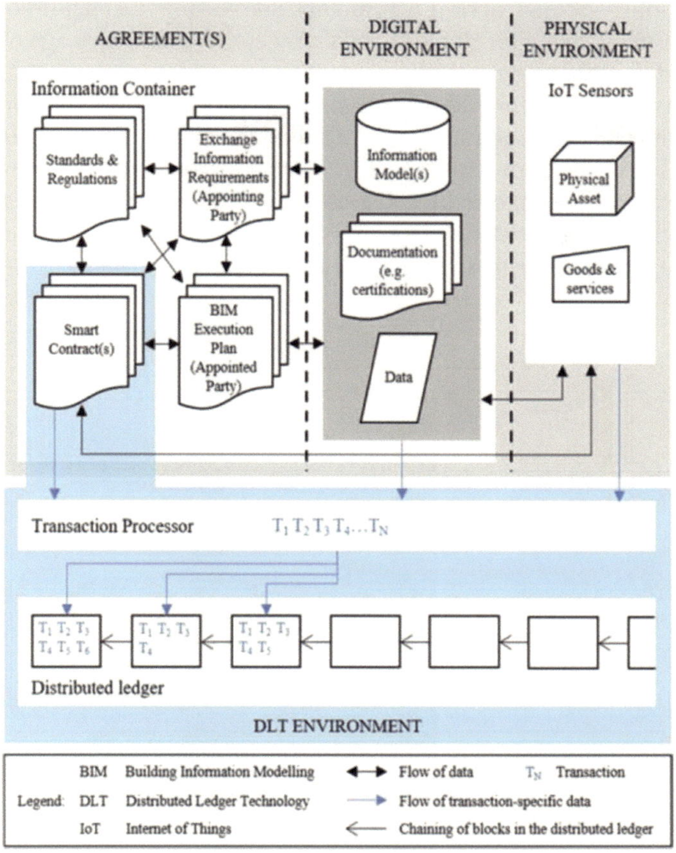

Figure 3.
Integration of BIM, IoT, iContract and Blockchain with the physical data from a construction project [41].

will be a substantial obstacle for the iContract to overcome. The degree to which they are truly unique is certainly open to debate as the industry evolves to a more manufacturing-based model and fact that all construction projects involve elements of repeatable processes that can be automated lends itself to automation.

5. Conclusion

Technological innovation is occurring at a growing pace as society has entered the digital age, and the construction industry is in a race to catch up with the digital capabilities of other sectors. Unfortunately, the construction industry has a historically short-sighted view on innovation with investment relying on immediate value. Knowing the antecedents of usefulness for any technology gives organisations the tools to present the case for adoption more effectively and convincingly. The iContract must present value to its users through the identification of the repeatable processes that it could be applied to. By the careful targeting of the technology to address the 'low-hanging fruit' problems of current contractual practices, the iContract can gain acceptance and begin to push the boundaries of digital automation into the contract process.

The notion that iContracts will be autonomously controlling construction projects based on data from advanced sensors acting as a certifying authority will not be achieved overnight. The advances in BIM, in multi-party contracts, in project insurance can all be seen as a pre-cursor for the type of paradigm shift required to achieve autonomous construction. Ultimately, addressing the current technological barriers is a waiting game for the iContract concept to reach the stage of maturity where it is indisputable from a legal perspective.

The iContract concept would bring enough disruption to reform contract practices within the construction industry and support its advancement into the digital revolution. This would allow the industry to better manage resources, reduce costs, reduce project durations and reduce dispute. As the iContract concept evolves, many of the challenges identified would be addressed and further opportunities will become apparent as trust in digitalisation increases.

Author details

Alan McNamara
University of New South Wales, Sydney, Australia

*Address all correspondence to: alan@icontract.ai

References

[1] Li J, Greenwood D, Kassem M. Blockchain in the built environment and construction industry: A systematic review, conceptual models and practical use cases. Automation in Construction. 2019;**102**:288-307

[2] Neely A. Carrying the Torch for Digital Built Britain. BIMPlus; 2018. [Online]. Available from: http://www.bimplus.co.uk/people/carrying-torch-digital-built-britain/?utm_source=dlvr.it&utm_medium=twitter. [Accessed: 18 June 2019]

[3] Farmer M. Modernise or Die: The Framer Review of the UK Construction Labour Market. London: Construction Leadership Council; 2016

[4] Woodhead R, Stephenson P, Morrey D. Digital construction: From point solutions to IoT ecosystem. Automation in Construction. 2018;**93**:35-46

[5] McKinsey. Reinventing Construction: A Route to Higher Productivity. McKinsey & Company; 2017

[6] Khan M, Wu X, Xu X, Dou W, editors. Big data challenges and opportunities in the hype of industry 4.0. Communications (ICC). In: 2017 IEEE International Conference on. IEEE; 2017

[7] Mason J. BIM fork: Are smart contracts in construction more likely to prosper with or without BIM? Journal of Legal Affairs and Dispute Resolution in Engineering and Construction. 2019;**11**(4):02519002

[8] McNamara A, Sepasgozar SM, editors. Barriers and drivers of Intelligent Contract implementation in construction. In: 42nd AUBEA Conference; 2018; Singapore. 2018

[9] Winfield M. Blockchain and smart contracts: Binary solutions in a non-binary industry? Planning Building and Control Today. 2018. [Online]. Available from: https://www.pbctoday.co.uk/news/bim-news/blockchain-and-smart-contracts-binary-solutions-in-a-non-binary-industry/39349/ [Accessed: 18 June 2019]

[10] Davis N, editor. What is the fourth industrial revolution. In: World Economic Forum. 2016. [Online]. Available from: https://www.weforum.org/agenda/2016/01/what-is-the-fourth-industrial-revolution/ [Accessed: 18 June 2019]

[11] Leviäkangas P, Mok Paik S, Moon S. Keeping up with the pace of digitization: The case of the Australian construction industry. Technology in Society. 2017;**50**:33-43

[12] Allen. Productivity in the Buildings Network: Assessing the Impacts of Building Information Models. Sydney: Built Environment Innovation and Industry Council; 2010

[13] ABS. Feature Article: A Statistical Overview of the Construction Industry: Australian Bureau of Statistics; 2010. Available from: http://www.abs.gov.au/AUSSTATS/abs@.nsf/Lookup/1350.0FeatureþArticle1Octþ2010

[14] Fitch T. Construction Industry Missing Out on Billions in R&D Tax Credits 2016. Available from: https://www.constructionnews.co.uk/analysis/expert-opinion/billions-in-tax-credits-are-being-missed-by-construction/10014472.article

[15] Schumpeter JA. Capitalism, Socialism and Democracy. New York: Routledge; 2010

[16] Schmidt R, Möhring M, Härting R-C, Reichstein C, Neumaier P, Jozinović P, editors. Industry 4.0-potentials for creating smart products: Empirical

research results. In: International Conference on Business Information Systems. Poznan: Springer; 2015

[17] Oesterreich TD, Teuteberg F. Understanding the implications of digitisation and automation in the context of Industry 4.0: A triangulation approach and elements of a research agenda for the construction industry. Computers in Industry. 2016;**83**:121-139

[18] Manyika J, Chui M, Madgavkar A, Lund S, editors. Technology, Jobs, and the Future of Work. McKinsey Global Institute Technology Briefing Note Prepared for the Fortune Vatican Forum; 2016

[19] Gartner. Gartner Identifies the Top 10 Strategic Technology Trends for 2014. 2013. Available from: https://www.gartner.com/newsroom/id/2603623

[20] McMalcolm J. How Big Data is transforming the construction industry. Construction Global. 2015. [Online]. Available from: https://www.constructionglobal.com/equipment-and-it/how-big-data-transforming-construction-industry [Accessed: 18 June 2019]

[21] Burger R. How the construction industry is using. Big Data. 2017. Available from: https://www.thebalancesmb.com/how-the-construction-industry-is-using-big-data-845322

[22] Peters GW, Panayi E. Understanding Modern Banking Ledgers through Blockchain Technologies: Future of Transaction Processing and Smart Contracts on the Internet of Money. In Banking Beyond Banks and Money. Cham: Springer; 2015. pp. 239-278

[23] Mason J. Intelligent contracts and the construction industry. Journal of Legal Affairs and Dispute Resolution in Engineering and Construction. 2017;**9**(3):04517012

[24] Wright A, De Filippi P. Decentralized Blockchain Technology and the Rise of Lex Cryptographia. 2015. Available from: https://ssrn.com/abstract=2580664

[25] Dakhli Z, Lafhaj Z, Mossman A. The potential of Blockchain in building construction. Buildings. 2019;**9**(4):77

[26] Manyika J. The Internet of Things: Mapping the Value beyond the Hype. McKinsey Global Institute; 2015

[27] Wu T. When code isn't law. Virginia Law Review. 2003;**89**(4):679-751

[28] De Filippi P, Hassan S. Blockchain technology as a regulatory technology: From code is law to law is code. First Monday. 2016;**21**(12) [Online]

[29] Ekblaw A, Azaria A, Halamka JD, Lippman A, editors. A case study for Blockchain in healthcare:"MedRec" prototype for electronic health records and medical research data. In: Proceedings of IEEE Open & Big Data Conference. 2016

[30] Guo Y, Liang C. Blockchain application and outlook in the banking industry. Financial Innovation. 2016;**2**(1):24

[31] Heywood M. Construction News—Smart Technology in Construction and the Legal Implications. 2016. [Online]. Available from: https://www.constructionnews.co.uk/archive/smart-technology-in-construction-and-the-legal-implications-29-11-2016/ [Accessed: 18 June 2019]

[32] Hughes D. The Impact of Blockchain Technology on the Construction Industry. 2017. [Online]. Available from: https://medium.com/the-basics-of-blockchain/

the-impact-of-blockchain-technology-on-the-construction-industry-85ab78c4aba6 [Accessed: 18 June 2019]

[33] Arcadis. Global Construction Disputes Report 2018: Does the Construction Industry Learn from Its Mistakes. 2018

[34] HMG. Digital Built Britain: Level 3 Building Information Modelling—Strategic Plan. London: HM Government; 2015

[35] Szabo N. Smart Contracts. 1994. Available from: http://szabo.best.vwh.net/smart.contracts.html

[36] Morgan P. Bitcoin South: Innovating Legal Systems through Blockchain Technology. 2014. [Online]. Available from: https://bravenewcoin.com/insights/pamela-morgan-at-bitcoin-south-innovating-legal-systems-through-blockchain-technology [Accessed: 18 June 2019]

[37] Norton-Rose-Fullbright. Smart Contracts: Coding the Fine Print—A Legal and Regulatory Guide. 2016. Available from: http://www.nortonrosefulbright.com/knowledge/publications/137955/smart-contracts-coding-the-fine-print

[38] Clack CD, Bakshi VA, Braine L. Smart contract templates: Foundations, design landscape and research directions. arXiv Preprint arXiv:160800771. 2016

[39] Mason J, Escott H, editors. Smart contracts in construction: Views and perceptions of stakeholders. In: Proceedings of FIG Conference; May 2018; Istanbul. 2018

[40] Li J, Greenwood D, Kassem M, editors. Blockchain in the Built Environment: Analysing Current Applications and Developing an Emergent Framework. Ljubljana: Diamond Congress Ltd; 2018

[41] Li J, Kassem M, Ciribini A, Bolpagni M, editors. A proposed approach integrating DLT, BIM, IoT and smart contracts: Demonstration using a simulated installation task. In: International Conference on Smart Infrastructure and Construction 2019 (ICSIC) Driving Data-informed Decision-making. Cambridge: ICE Publishing; 2019

[42] Latham M. Constructing the Team: Joint Review of Procurement and Contractual Arrangements in the United Kingdom Construction Industry. London: HMSO; 1994

[43] Wang D, Fan J, Fu H, Zhang B. Research on optimization of big data construction engineering quality management based on RnN-LSTM. Complexity. 2018;**2018**:1-16

Chapter 8

5D BIM Applications in Quantity Surveying: Dynamo and 3D Printing Technologies

Anqi Shi, Sara Shirowzhan, Samad M.E. Sepasgozar and Alireza Kaboli

Abstract

Digital construction is increasingly introduced to the architecture, engineering, and construction (AEC) industry. The fifth dimension of Building Information Modeling (5D BIM) has a significant contribution to the AEC industry, such as managing time and costs and resources management. However, 5D BIM has not fulfilled its promise completely. The major challenge of implementing the 5D BIM technology and applications is the interoperability between technologies and the low accuracy of measurement estimations. Dynamo is adopted in this chapter for detailed quantity measurement. Three-dimensional printing (3DP) is one of the additive manufacturing technologies which is recommended to be used in construction. 3DP is aiming to create complex and customized geometries. This chapter demonstrates how to apply the Dynamo scripts of detailed quantity take-off for estimating the volume of elements created by 3DP. Dynamo as a quantity take-off Add-in in Revit is always used for detailed quantity take-off or precise model created. After the detailed quantity take-off from Revit, the data can be exported into Excel extension, which allows the quantity surveyors to insert the price per unit to the file and to generate a Bill of Quantity (BOQ). This chapter will offer a procedure for applying Dynamo through two selected case studies.

Keywords: 5D BIM, quantity surveyor, dynamo, Revit, decision-making, quantity extraction, 3DP, data-driven procedure

1. Introduction

Building Information Modeling (BIM) has been adopted in the AEC industry for a long time, especially in the offsite (prefabrication) construction [1, 2]. 5D BIM is Building Information Modeling that includes a 3D model plus scheduling (4D) and cost management (5D) [3]. 5D BIM is an advanced technology in the AEC industry, which can be used for managing time, cost, and resources; it can even handle the logistical site plan. 5D BIM can help quantity surveyors review alternative designs during the early stage of the project as a decision-making tool [4], since the 5D BIM can quickly extract approximated quantity from different 3D BIM models and then add the 4D schedule to finalise the 5D cost budget. However, the development of 5D BIM is hampered [3], by many reasons, such as the high initial implementation costs [5] but low accuracy [3]. Moreover, the current 5D BIM software and applications cannot fulfil the subcontractors' requirements which are focused on detailed works, such as external cladding. They are appropriate for the rough cost estimation of a large scope, which is satisfied with the requirements from project planners and head contractors. The quantity take-off of current major 5D BIM applications or software is approximate quantity instead of exact quantity due to the Level of Development (LOD) limitation. Dynamo is more appropriate for the detailed quantity take-off, such as three-dimensional printing (3DP) elements. 3DP is used to print precise and complex elements [6].

Hall and Tewdwr-Jones [7] demonstrated that the lack of cooperation and difficulty of information sharing are the major reasons for the communication issue during the entire project life cycle. However, the BIM cannot always guarantee the communication to be effective [5]. The chapter will investigate how Dynamo Revit can be used for detailed quantity take-off and how to analyse the combination of 5D BIM quantity take-off appliances (Dynamo) with 3DP. A face-to-face semi-structured interview with regard to the case study will be conducted. Scopus database was the main source used for literature review in this chapter, since it has a relatively wide range of data [8], especially for BIM and the 3DP.

Figure 1 visualizes the geographical locations of the 5D BIM papers published in different countries or territories. The nodes present the number of papers that contributed to each country or territory. These nodes are grouped into two groups with different colors by running the Modularity program. The Modularity program implements a community detection algorithm, called the Louvain method [9].

2. Aims and objectives

The aim of this chapter is to present the current status of 5D BIM and the prospect of the integration of 3DP with 5D BIM in the AEC industry. The key objectives of these investigations are as follows:

Objective 1: To identify the advantages and challenges of the 5D BIM implementation in the AEC industry.
Objective 2: To evaluate the implementation of 3DP in the AEC industry.
Objective 3: To develop a 5D BIM-3DP integrated workflow for the project's cost analysis.

Figure 1.
Geographical locations of articles contribution.

3. Literature review

3.1 BIM

'A BIM is a shared knowledge resource for information about a facility forming a reliable basis for decisions during its life cycle; defined as existing from earliest conception to demolition' [10]. Regarding BIM, 4D BIM is about scheduling, 5D BIM is used for the cost analysis, 6D BIM is for sustainability analysis, and 7D is about the facility management [11]. BIM generates a comprehensive environment for a project and also assigns personnel responsibilities [12]. The AEC industry is always low in productivity and lacks cooperation and innovation. BIM has the potential to solve these challenges [13, 14].

3.2 5D BIM implementation advantages

As a useful information platform, 5D BIM is not only utilized during the project construction process but also during the entire life cycle. For example, BIM can share and update the drawings and specifications easily in the cloud database, and then 5D BIM can generate more consistent and accurate cost estimation. BIM platform provides a smooth flow of information sharing among stakeholders to transfer the information quicker and easier among multidisciplines [15] to reduce errors or unnecessary works. Additionally, 5D BIM can show a clear budget and construction progresses to participants [3]. Moreover, 5D BIM is time efficient for alternative design analysis and decisions at the early stage. 5D BIM often conducted the cost management and cost analysis in other software or application. 5D BIM can better monitor the project costs not only in the short term but also in the long term by including related information and resources during [16].

3.3 5D BIM implementation challenges

The challenges of 5D BIM adoption are, for instance, software securing, training investment, and low time efficiency, lacking standards of software compatibility [16]. Sattineni and Macdonald [3] reported that 5D BIM always has high initial installation costs and requires new expertise, such as computer experts, compared to traditional construction. Thus, the company is unwilling to change [3]. The other limitation of 5D BIM is the low LOD, which cannot extract the detailed data from the 3D model for cost estimation. Aibinu and Venkatesh [5] stated there are several difficulties of 5D BIM used by the quantity surveying organization, such as the low LOD of a model. This statement is based on 40 responses received from their 180 QS web survey, and two in-depth interviews are conducted. Therefore, due to the lack of detailed information, the team always spent longer time to make sure the quantity take-off is accurate [3, 5].

BIM-based clash detection is a quantity take-off method, which was introduced by Khosakitchalert et al. [4] to enhance the level of the quantity take-off accuracy of the layered structure. BIM-based clash detection approach extracts quantity information by geometry data of the model, such as the area and the name of the target objects [4]. Dynamo-extension is one Add-in of the Revit, and it calculates the quantity for the components by script [4]. Khosakitchalert et al. [4] stated that the LOD of the layered structures is low, such as walls, which causes low accuracy of the quantity take-off [4]. Different layers of the components have different dimensions due to the overlapped connection requirements [4]. Revit supports two different quantity take-offs, one is the material take-off and the other one is the quantity take-off [4]. The quantity take-off is based on elements, and the material take-off is based on material information [4]. The data or 3D model can be processed by creating an algorithm in Dynamo [4]. Dynamo cannot conduct quantity take-off for one element when various materials are used in this element [4]. Besides, there is rare software available for higher-level BIM functions at the same time [4]. The common 5D BIM or BIM software is not suitable for the detailed quantity take-off, for instance, cladding tasks. Since architectures and designers always add many details into the drawings instead of models, most of the subcontractor companies used BIM only for the 3D model visualization. Some subcontractors do not share the information with the crew members timely, and the designers might not be willing to share the models with the construction team. It is easy to case unneeded work due to inefficient communication and inconsistent drawings. In addition, the design drawings and specifications may conflict, which is caused by the errors or data missing and ultimately leads to inaccurate cost estimation [17, 18]. Furthermore, the model's errors may be not revised timely, which will affect the accuracy of the quantity take-off [3]. Mayouf et al. [17] also stated that the 5D BIM implementation requires a comprehensive workflow and information. One of the difficulties of quantity surveyors' work is the poor quality of design documentation. Multidisciplinary and collaborative approaches play an essential role in improving the efficiency of communication and the quality of the documentations [17].

The BIM models are not designed for the quantity take-off or the cost estimation, so some details will not be shown in the models. It is a significant issue for the quantity surveying companies [5]. Thus, it is tough to extract the detailed quantity from elements for quantity estimation. 5D BIM provides timely communication but cannot guarantee the quality of the communication. At the end of 2016, Software Advice (UK) analysed and reported that 50% of the investigated small and medium enterprises still conduct the cost estimation manually [16]. There are many limitations of CUBIT adoption in New Zealand and Australia, such as low accuracy of quantity take-off or detailed quantity missing [16, 19]. CUBIT Buildsoft is developed by MiTek (an Australian company) and used in Australia, New Zealand, the United Kingdom, and Ireland [16].

There are other challenges of 5D BIM applied in a project, such as the standardization [20] and high training costs that some organizations may not able to afford [20].

Different from academia, the industry believes BIM needs a process change approach than information—drive approach [17]. **Table 1** shows the academic paper reviewing the possibility of integrating Dynamo with other tools. These papers were selected from the Scopus database since they focus on the Dynamo Revit. The Dynamo is the essential step for the 5D BIM-3DP integrated workflow. **Table 2** presents those papers which focused on sustainability applications. In both tables, green color means the article is related to Dynamo and Revit for the quantity take-off; blue color means partially related.

3.4 3D printing

Contour crafting (CC) was developed by researchers at Loughborough University, UK [38, 39]. The conventional construction methods damage the environment [40] with lower efficiency, and it is hard to achieve a project on time and within budget without compromising the quality [41, 42]. 3DP can reduce the cost, shorten construction duration, and minimize waste. At the same time, 3DP is efficient and sustainable and can achieve customization [43]. Additionally, Tay et al. [44] reviewed 115 relevant articles in the Web of Science and Science Direct and stated

Aims and objectives	Method and case studies	Limitations or suggestions for future	Findings or contribution
To develop a BIM-based impact of change order evaluation systems [21]	System (Autodesk Revit, Dynamo, Microsoft Excel, and VBA) quantifies three impacts of change in projects, which are physical conditions, schedule, and budget	Limited to an overview of system architecture and each component of the system	Dynamo can import the data to Revit and then export it to other parts of the system. Besides, use Dynamo to store the cost and schedule of the 3D model
To utilize the development of BIM technology in bridge engineering [22]	Case study: Shenyang Wuai Overpass is selected	Not mentioned	Through Dynamo, design heteromorphic ramp parameters with high accuracy
To analyze and evaluate the BIM statement in a heritage building and further development [23]	Case study: Jewel Tower, UK, is selected	Potential of reporting and forecasting complex models should be examined	Dynamo (Python - algorithm) used to portray the spatial distribution of moisture with RGB parametrization
To utilize more tools for integrated digital simulation [24]	Case studies from two companies: Danish architecture firm BIG and international engineers BuroHappold	Not mentioned	Using the "Dynamo Masterplanning Tool" for different mechanical, electrical, and public health based on building form parameters. Also, it can be used to update the floor area and the external wall area
Capture facility information to deliver integrated handover deliverables [25]	Case study: Two-storey educational building in Rocky Mountain Region, USA	The framework needs customization for each project based on the owner and end-user needs	Dynamo workflow used to fulfill the parameters of the 'Master Format Division' with the appropriate value in the Building Handover Information Model (BHIM) framework
To utilize interdisciplinary and overall digital design methods [26]	Case studies such as structural BIM models for the 'HENN Architekten', a competition, Shenzhen, China, 2013	Not mentioned	Grasshopper for Rhino or Dynamo for Revit can visualize and assess the designs and generate many alternative designs in a short time
Through Dynamo, visual programming language, based on the parameters, combined the bird threat assessment of facade material, analysis building geometry relative to the materials, and evaluate the users' input to the building operation [27]	The resulted workflow allows designers to start building a model in Revit using custom families, which contain the factors and then run the Dynamo	The problem of walls and windows as they cross over the two zones in Dynamo can be solved by Dynamo, but difficult	Bird collisions on building facades are important and should through Dynamo/Revit (VPL/BIM) be revised and released, to present a standard way to assess compliance with LEED Pilot Credit 55

Aims and objectives	Method and case studies	Limitations or suggestions for future	Findings or contribution
To develop a new mesh-to-HBIM modeling workflow and connect the elements of HBIM and historical knowledge through integrating BIM management system [28]	Case study: St-Pierre-le-Jeune Church, Strasbourg, France	Not mentioned	Dynamo can provide the interactive interface in the study, and users can browse the semantic information reserved in the ontology database and the 3D model in BIM at the same time
To present a workflow that incorporates with Revit of an architectural precast concrete manufacturer [29]	Case study: A hospital designed by Flad Architects which expands the University of Florida Health Shands complex, Gainesville, Florida	In the future, the workflow could involve the subcontractors early for the advice and decision-makings. It could be called as 'Design-Assist'	Dynamo can create any of the panels of the case study of an architectural precast concrete manufacturer that the Revit model could
To reduce labour involvement during the modeling process by shaping the mesh geometry [30]	An antique monument, Petit château du Meisenbach, and a church of the Abbey of Niedermunster	Dynamo can be used to create a building modeling based on the 'family' of objects	Dynamo can develop a semiautomated function to reduce the human process. Dynamo can adjust the element parameters by parameter name and value
To have the BIM model as a control system of building energy performance service [31]	Case study: Building facade design	Not mentioned	Dynamo is used for parametric Adaptive Skin System (PASS), which consists of kinetic facade components
To extract and process the BIM data through Revit, the Dynamo can help for data processing and analyzing [32]		There are no ready modes for all functions of the Revit API	Dynamo can develop external applications that provide extensive opportunities for expansion of applications, convenience, speed, and, as a consequence, productivity in Revit

Table 1.
Summary of papers examines in dynamo and Revit applications.

3DP could achieve higher productivity with lower investment cost and waste. Tay et al. [44] also indicated that the 3DP is appropriate for flexible component customization. Sakin and Kiroglu [45] stated the 3DP might be developed along with the traditional construction method.

However, the 3DP is also not fulfilled its promise completely [46]. Although the 3DP can reduce the labour to save the costs, the more equipment is required in the construction site, and the depreciation costs also need to be considered. Therefore, it is hard to state if the 3DP is saving money or not. The material technology is a significant challenge for 3DP development. Gosselin et al. [47] are against to the idea that the 3DP can achieve the feasibility 'free-form' structures. For instance, if 3DP material is clay, then the suspended parts cannot stand immediately after the printing without any support (see **Figure 2**). **Figure 2** shows the sagging section

Aims and objectives	Method and case studies	Limitations or suggestions for future	Findings or contribution
Develop an integration interface of manufacturer-based LCA data in BIM by combining Revit, Dynamo, and Excel [33]	Case study: The scientific approach for decision-making, with models. Single-family social housing unit model, municipality of Acailandia, state of Maranhao, Brazil	The complexity of the programming; it is difficult to import and extract the data from different construction subsystems	Obtain environmental performance for decision-making in the initial design stages in an automated way of information insertion/ extraction to or from the model
To encourage the BIM for the design uncertainties affecting building energy performance [34]	Case study: Residential unit in College Station, Texas, USA	Not mentioned	Dynamo exports input variables to a spreadsheet-based energy analysis tool for uncertainty and sensitivity analyses
To investigate the feasibility of connecting environmental sensors such as light, humidity, or CO_2 receptors to a BIM [35]	Case study	The slower interaction time in Dynamo Revit than Grasshopper Rhino	Dynamo and Revit API are interfaces of case studies, the software linking the environment sensor with BIM, such as linking the Revit and the Arduino board. Dynamo also can help to adjust the object's parameters
To create a smarter and more flexible BIM by programming and scripting [36]	Case study: Facade component	Verify the level of customization and flexibility with basic programming knowledge	Dynamo is the major concept in the case of a solar-activated kinetic façade component through creating programming
To extend Dynamo by using a building energy simulation package, controlled by a virtual model's response through light level sensors [37]	Case studies used for Energy Analysis of Dynamo linking photoresistor value to 3D model and produce dynamic solar shades	Lack of consistent nomenclature between Revit and Dynamo about families and parameters; less stability and missing nodes and features	Dynamo can use with other software programs, although currently it is not widely implemented in practice. Test interactively updating shading components for a building facade based on solar angles

Table 2.
Summary of papers focusing on sustainability issues by implementing dynamo and Revit applications.

(red square) of a clay 3DP model. The 3DP challenges are summarized into (1) material development, (2) reinforcement implementation, and (3) process parameter optimization, such as the flow rate of material and print speed [44]. Since the materials are the major challenge of 3DP development, the 5D BIM-3DP integrated workflow in this chapter focuses on the material quantity take-off.

3.5 BIM-3DP integration

There are many projects adopted by BIM technologies to combinate with offsite construction. Offsite construction is also named prefabrication construction, which is a method to produce the standardized components under the controlled

Figure 2.
Column models in laboratory scale with 30 degrees.

environment to ensure the project's quality and efficiency [48]. 3DP is very similar to offsite construction. Although 3DP lacks the corresponding regulations and standards, and the data exchange efficiency might be improved after combinating with the BIM, 3DP is time efficient and sustainable and can achieve customization [43] than the offsite construction. Thus, BIM and also 5D BIM can integrate with 3DP instead of offsite construction. BIM can automate the 3DP since it can store the equipment and manufacturing information, such as printer control data [44]. The integration of the multi-nozzle with a hybrid 3DP system is useful for concrete structures with various materials and elements, such as rebar. The combination of 3DP with BIM can monitor the variables in the construction site [49, 50]. Also, this 5D BIM-3DP integrated workflow can achieve a simple customized building structure [44] in a laboratory scale and be brought to the construction site or meeting for communication quality improvement.

4. Methodology

The methods in this section are two case studies. Case study 1 is aiming to apply the 5D BIM quantity take-off to the multilayer wall to get all the layers' quantity information at the same time. The second case study is to apply 5D BIM to the 3DP elements. The builder and stakeholders cannot get all the layers' quantity information at the same time by using the conventional 5D BIM applications or software. This chapter adopts the Dynamo script principle that is mentioned in Khosakitchalert et al. [4], to design a script for all the layers' quantity take-off from one multilayer wall at the same time, to detect the meticulous differences among the layers. Furthermore, this script can be used to extract the detailed quantities from the 3DP elements.

Figure 3 is only a simple script to show the area quantity extracted from the existing models. Also, Dynamo can be used to create a precise 3D model by node (units in Dynamo script) (see **Figure 4**). Later, the 3D model's dimensions can be revised or changed precisely through the nodes in the core category of the Dynamo library. Each node is a function or order of the executed commands [4].

The floor is just an example to show the basic concept for the entire programming.

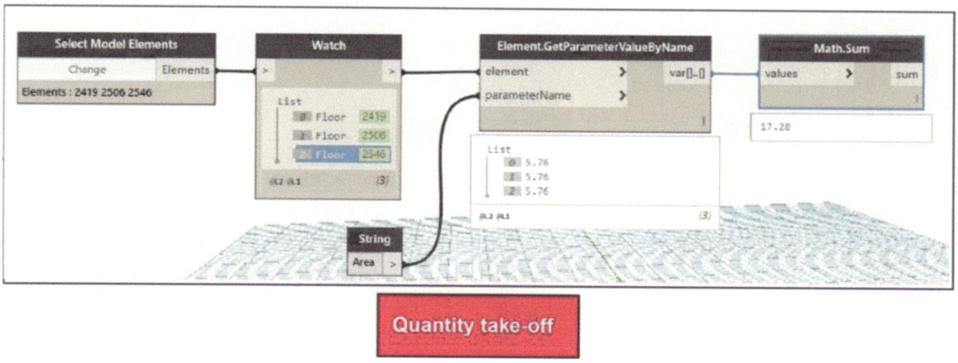

Figure 3.
Dynamo-extension in the Revit for detailed quantity take-off.

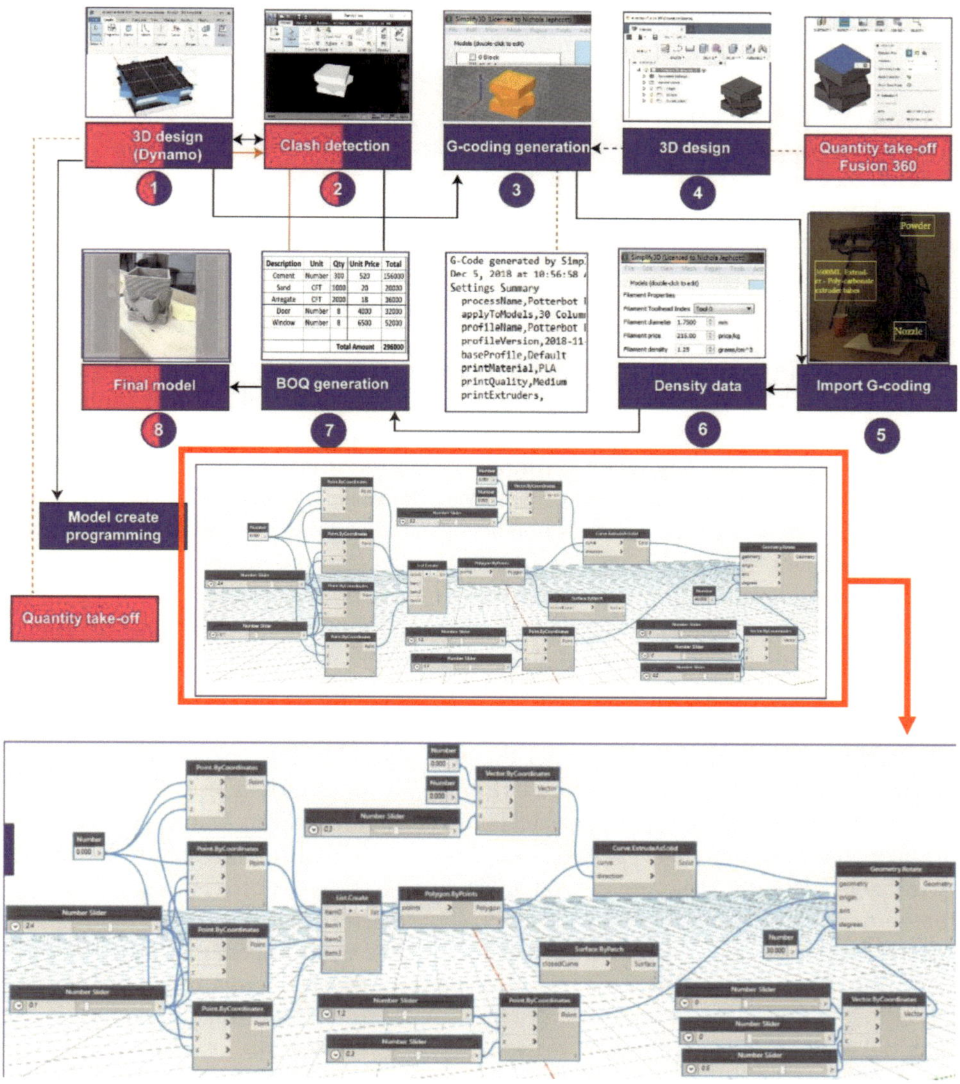

Figure 4.
Dynamo and fusion quantity take-off.

The Dynamo for element quantity take-off programming consists of six stages:

1. Select the 'Select Model Elements' node from the Dynamo library, and select the target element in the 3D model.

2. Select the 'Watch' node to list the detail code of each item.

3. Connect the 'Watch' node to the 'Element. Get Parameter Value By Name' node. The results listed below are the area (parameter name) of each visible element.

4. In this scenario, the value or the outcome of the parameter name is area.

5. Select the 'Math Sum' node to calculate the total area of the 3D model.

6. This programming is also suitable for the users to gain other material information, such as the type of material.

5. Case studies

Two case studies are contained in this section. One is applied this detailed quantity take-off to one target multilayer wall in the facade case study; the other one is regarding the combination of 5D BIM-3DP integrated workflow.

5.1 Case study 1: multilayer wall quantity take-off

In the facade installation project, there is an external wall consisting of different layers such as cladding, sub-framing, waterproofing membrane, stud-framing, insulation, and plasterboard. However, the area quantity (m2) of these layers cannot be the same, especially when it connects with other walls. The accuracy of the element's quantity take-off in this project may decrease by the LOD issue [4]. For instance, the quantity take-off could be accurate for doors, windows, or structural framing, but not the external cladding, or internal plasterboard, where there are many details. **Figure 5** shows different wall types at two different levels.

The highlighted area is wall type 8.25, and according to the wall type schedule document, it consists of external cladding, sub-framing, waterproofing membrane, stud wall, insulation, and plasterboard. The Dynamo script can help to extract the element quantity. For example, if quantity surveyors allow that the area of wall type 8.25 is 500, all the other layers of this wall should be 500 m2 as well (**Figure 5a**). However, wall type 9.03a (**Figure 5b**) on the right side of wall type 8.25 is a concrete wall. It has the plasterboard inside. However, wall type 8.16 (**Figure 5b**) is behind wall type 9.03a (and perpendicular to it), which reduced the quantity of plasterboard on wall type 9.03a compared to the other layers. The conventional 5D BIM estimating software, such as Cubit, cannot detect these detail differences. Thus, the quantity information of some walls' layers might be slightly inaccurate if we use conventional 5D BIM software to get the quantities. Although it is helpful to use BIM for estimating as it reduces time and human mistakes, it is essential to consider its limitation for some specific works. The Dynamo in this case study can detect a small difference between the wall layers. The quantity take-off for each layer would be more accurate than the conventional 5D BIM estimating software.

Figure 6 is shown how to use the Dynamo script to extract the detailed quantity of a multilayer wall. Users can add extra nodes from the Dynamo library for other

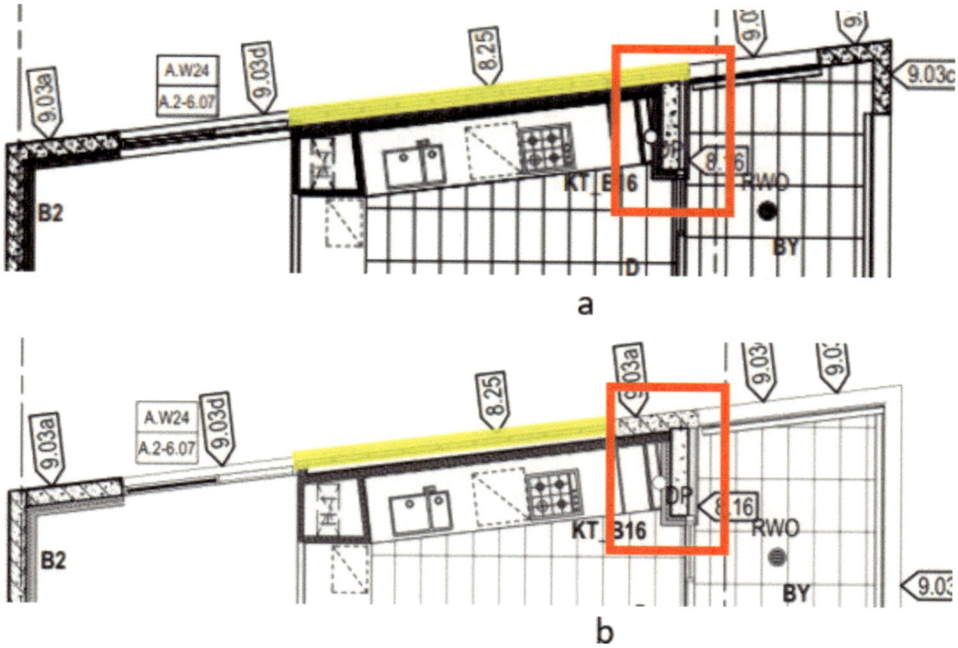

Figure 5.
Detail plan and section drawings. (a) Planning drawing levels 02, 04, and 06. (b) Planning drawing levels 03 and 05.

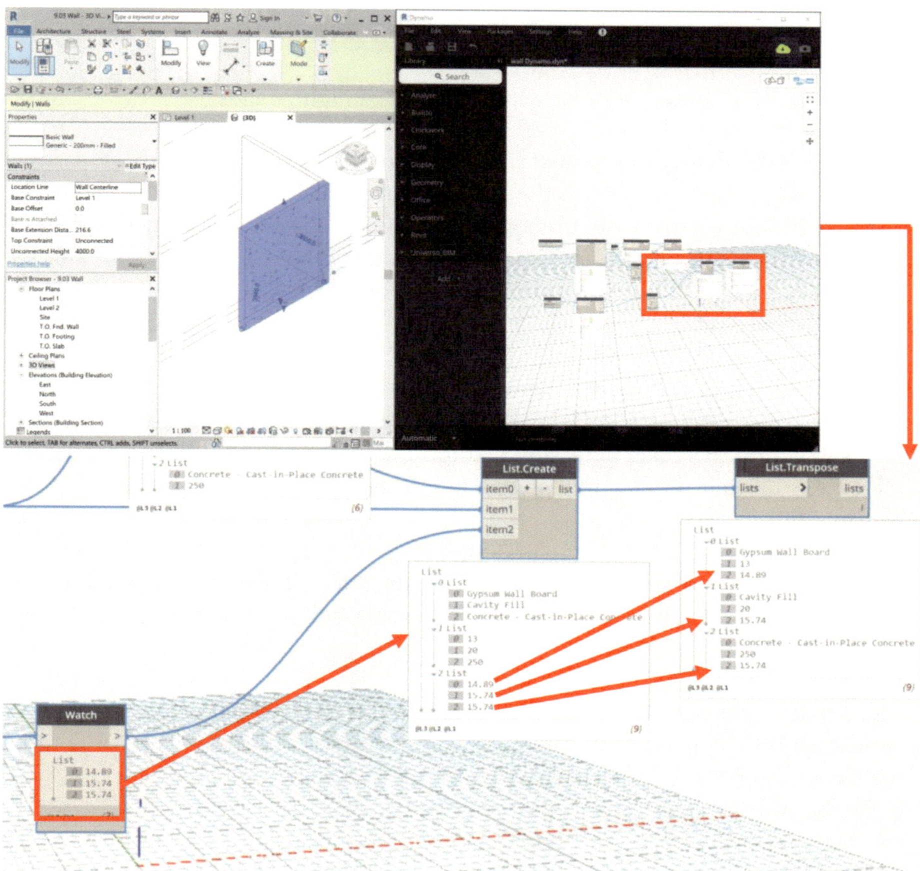

Figure 6.
The quantity take-off for multiple layers of the wall through dynamo.

purposes. This Dynamo script was developed based on the script in **Figure 3**. The updated Dynamo script can list all the target layers' information of multiple walls at the same time, including the name, width, and area of each layer. The rest of the script is the same with **Figure 3**. **Figure 6** shows that the area of each layer's quantity is slightly different: the gypsum wallboard is 14.89 m^2, but the other two are 15.74 m^2. It states the Dynamo script can be accurate on two decimal places for the quantity estimation. For instance, in this case study, the differences between 14.89 m^2 and 15.74 m^2 are 0.85 m^2. **Figure 6** is only shown one multilayer wall's Dynamo script. The users can create as many as groups they want to check the entire project walls' layer information at the same time. The quantity data can be exported into an Excel extension, which allows the quantity surveyors to insert the price per unit to the file and to generate a Bill of Quantity (BOQ).

5.2 Case study 2: 5D BIM-3DP workflow

Research designed a workflow that combines the 5D BIM with 3DP by Dynamo. As mentioned above, the 3DP might be applied to tradition construction [45]. The 3DP technologies are advance on small-size components, which requires an accurate quantity take-off. This 5D BIM-3DP integrated workflow can be used to estimate the 3DP elements' quantities, to improve the accuracy of the cost estimation. In addition, the 3DP can print scaled-down models of some special parts to improve communication efficiency. The Dynamo script for the quantity take-off is based on the script in **Figures 3** and **6**. Sometimes, the LOD of the Revit architecture model is not enough to create the 3D model. The users also can use the Dynamo script to create a model (see **Figure 4**) and then measure it.

Figure 4 demonstrated the integration of the 5D BIM (Dynamo for quantity take-off) with the 3DP.

The 5D BIM-3DP integrated workflow consists of eight stages:

1. Create a 3DP model using Dynamo (see **Figure 4**) or Fusion 360 (Step 4). Fusion 360 is more suitable in the manufacturing industry, which is appropriate for accurate quantity take-off [51]. The Dynamo script in **Figure 4** is a process for model creation since the quantity take-off script is the same as in **Figures 3** and **6**. The model should be imported into Navisworks by '.rvt' or '.rfa' field extension.

2. Conduct the clash detection in Navisworks; 4D schedule information in the Navisworks is added into BOQ [52]. Then, the users can revise the model in Revit based on the clash detection results. The 3D model should be saved as '. stl' extension in Revit (by installing a plugin—STL Exporter [53]) and Fusion 360.

3. Input the 3D model into Simply3D.*STL* extension, and the model would be sliced into the 2D printable layers; the G-coding will be generated automatically.

4. The models also can be created by Fusion 360 as mentioned above, and then repeat step 3 and make sure the models are ready to be printed. The quantity information of the models also can be found in Fusion 360. However, this software is more suitable for manufacturing industry's models.

5. Import the G-coding into the 3D printer by USB, and the final models can be printed.

6. The schedule, and other cost information, such as the density of the ink materials, is collected.

7. Generate and finalize a BOQ of 3DP samples based on the quantity take-off results from the Dynamo (the same steps as in **Figures 3** and **6**) and other cost information.

8. Final physical models in the desired scale are printed.

6. Findings

1. Conventional 5D BIM software and applications cannot extract the small element's quantities accurately at the same time. 5D BIM-3DP integrated workflow has been introduced based on the nature of 5D BIM and 3DP. The users can use the Dynamo to extract accurate quantities from small and precise elements at the same time. For instance, in case study 1—Multilayer wall quantity take-off—the Dynamo script can list the precise quantity of each layer of the multilayer wall at the same time. It is easier for the quantity surveyors, builders, and contractors to understand and check the construction details and then improve the accuracy of the cost estimation and project quality. Also, case study 2—5D BIM-3DP workflow—demonstrated this workflow also can be applied to the 3DP components.

2. Also, the Dynamo can help the users grab the other information quickly, such as the type of material. It helps the BOQ preparation and decision-making at the early stage of the project.

3. In the AEC industry, communication misunderstanding is a significant issue. It is because everyone is thinking in different ways. This 5D BIM-3DP integrated workflow can also be used to print and estimate the scaled-down or laboratory scale models for the complicated or specific components as the communication tool in the construction site or meeting to enhance the communication efficiency.

7. Conclusion

This chapter summarized the advantages and challenges of 5D BIM in the AEC industry. Also, the current conventional 5D BIM applications and software are not suitable for the detailed quantity take-off. The Dynamo, which is one of the Revit Add-ins, can help the users to extract the accurate quantity take-off. For instance, the Dynamo script in case study 1 lists the precise quantities of each layer of the multilayer wall at the same time. In addition, the status of the 3DP in the AEC industry had been evaluated. Since the 3DP has the potential to develop with the traditional construction industry, and 3DP in advance on the small and precise elements printing, the accuracy quantity take-off is very essential. The Dynamo also can be used to create and estimate the 3DP elements and components. The 5D BIM-3DP integrated workflow introduced in this chapter can help the quantity surveyors to get more accuracy quantity take-off for the 3DP. Moreover, the 5D BIM-3DP integrated workflow has the potential to reduce errors and misunderstanding during the communication. Besides, the highly accurate estimation of a 3D model at the early stage of the project would help the users to choose the best alternative design.

The 3DP scaled-down samples can be used to deliver and exchange the participant's idea during the meeting or in the construction site, which can improve communication efficiency. This 5D BIM-3DP integrated workflow is mainly focused on material quantity. Although the 5D BIM-3DP integration can reduce the labour to save the costs, more equipment is required in the construction site and the depreciation costs also need to be considered. Thus, the cost pattern may be changed. Thus, the other costs, such as equipment and labour salary, should be considered and investigated in the future as well.

Author details

Anqi Shi[1]*, Sara Shirowzhan[1], Samad M.E. Sepasgozar[1] and Alireza Kaboli[2]

1 Faculty of Built Environment, University of New South Wales, Sydney, Australia

2 Australian Cladding Specialists, Sydney, Australia

*Address all correspondence to: anqi.shi@unsw.edu.au

References

[1] Yin X, Liu H, Chen Y, Al-Hussein M. Building information modelling for off-site construction: Review and future directions. Automation in Construction. 2019;**101**:72-91

[2] Li CZ, Hong J, Xue F, Shen GQ, Xu X, Mok MK. Schedule risks in prefabrication housing production in Hong Kong: A social network analysis. Journal of Cleaner Production. 2016; **134**:482-494

[3] Sattineni A, Macdonald JA. 5D-BIM: A case study of an implementation strategy in the construction industy. In: ISARC. Proceedings of the International Symposium on Automation and Robotics in Construction, Vol. 31. IAARC Publications; 2014. p. 1

[4] Khosakitchalert C, Yabuki N, Fukuda T. The accuracy enhancement of architectural walls quantity takeoff for schematic BIM models. In: ISARC Proceedings of the International Symposium on Automation and Robotics in Construction, Vol. 35. IAARC Publications; 2018. pp. 1-8

[5] Aibinu A, Venkatesh S. Status of BIM adoption and the BIM experience of cost consultants in Australia. Journal of Professional Issues in Engineering Education and Practice. 2013;**140**(3): 04013021

[6] Ma GW, Wang L, Ju Y. State-of-the-art of 3D printing technology of cementitious material—An emerging technique for construction (in English). Science China-Technological Sciences. 2018;**61**(4):475-495. DOI: 10.1007/ s11431-016-9077-7

[7] Hall P, Tewdwr-Jones M. Urban and Regional Planning. Routledge; 2010

[8] Mongeon P, Paul-Hus A. The journal coverage of web of science and scopus: A comparative analysis. Scientometrics. 2016;**106**(1):213-228

[9] Start TQ. Gephi Tutorial Quick Start [Online]. 2010. Available from: https:// gephi.org/tutorials/gephi-tutorial-quick_start.pdf

[10] Moynihan GP, Harsh C. Current state of construction project delivery methods. In: IIE Annual Conference. Proceedings. Institute of Industrial and Systems Engineers (IISE); 2015. p. 174

[11] A. A. Digital. "05. All the BIM Dimensions. Available from: http:// www.axdstudio.com/bim-dimensions/ [Accessed: 10 September 2019]

[12] Ding Z, Liu S, Liao L, Zhang L. A digital construction framework integrating building information modeling and reverse engineering technologies for renovation projects. Automation in Construction. 2019;**102**: 45-58

[13] Arayici Y, Egbu C, Coates S. Building information modelling (BIM) implementation and remote construction projects: Issues, challenges, and critiques. Journal of Information Technology in Construction. 2012;**17**: 75-92

[14] Elmualim A, Gilder J. BIM: Innovation in design management, influence and challenges of implementation. Architecture, Engineering & Design. 2014;**10**(3–4): 183-199

[15] Grilo A, Jardim-Goncalves R. Value proposition on interoperability of BIM and collaborative working environments. Automation in Construction. 2010;**19**(5):522-530

[16] Mesároš P, Smetanková J, Mandičák T. The fifth dimension of BIM–implementation survey. IOP Conference Series: Earth and Environmental Science. 2019;**222**(1): 012003

[17] Mayouf M, Gerges M, Cox S. 5D BIM: An investigation into the integration of quantity surveyors within the BIM process. Journal of Engineering Design and Technology. 2019;**17**(3): 537-553

[18] Wu S, Ginige K, Wood G, Jong SW. How can building information modelling (BIM) support the new rules of measurement (NRM1). Royal Institution of Chartered Surveyors Research Paper; 2014

[19] Harrison C, Thurnell D. BIM implementation in a New Zealand consulting quantity surveying practice. International Journal of Construction Supply Chain Management. 2015;**5**(1): 1-15

[20] Smith P. Project cost management with 5D BIM. Procedia-Social and Behavioral Sciences. 2016;**226**:193-200

[21] Likhitruangsilp V, Handayani TN, Ioannou PG, Yabuki N. A BIM-enabled system for evaluating impacts of construction change orders. In: Construction Research Congress 2017; 2018. pp. 622-631

[22] Song FC, Xie LB, Li MC, Liu JT, Zhang GQ. Collaborative design of complex overpass based on BIM technology (in Chinese). Shenyang Gongye Daxue Xuebao. 2018;**40**(6): 692-696. DOI: 10.7688/j.issn.1000-1646. 2018.06.17

[23] Pocobelli DP, Boehm J, Bryan P, Still J, Grau-Bové J. Building information models for monitoring and simulation data in heritage buildings. International Archives of the Photogrammetry, Remote Sensing and Spatial Information Sciences-ISPRS Archives. 2018:909-916

[24] Peters B. Defining environments: Understanding architectural performance through modelling, simulation and visualisation (in English). Architectural Design. 2018; **88**(1):82-91. DOI: 10.1002/ad.2262

[25] Sadeghi M, Elliott JW, Porro N, Strong K. Developing building information models (BIM) for building handover, operation and maintenance (in English). Journal of Facilities Management. 2019;**17**(3):301-316. DOI: 10.1108/JFM-04-2018-0029

[26] May R. Engineering urban complexity: Bespoke integrated design (in English). Architectural Design. 2015; **85**(6):104-109. DOI: 10.1002/ad.1986

[27] Kensek K, Ding Y, Longcore T. Green building and biodiversity: Facilitating bird friendly design with building information models (in English). Journal of Green Building. 2016;**11**(2):116-130. DOI: 10.3992/ jgb.11.2.116.1

[28] Yang X, Lu YC, Murtiyoso A, Koehl M, Grussenmeyer P. HBIM modeling from the surface mesh and its extended capability of knowledge representation (in English). ISPRS International Journal of Geo-Information. 2019;**8**(7):301. DOI: 10.3390/ijgi8070301

[29] Collins J. Incorporating BIM into architectural precast concrete fabrication. ISARC. Proceedings of the International Symposium on Automation and Robotics in Construction. IAARC Publications; 2016. p. 1

[30] Yang X, Koehl M, Grussenmeyer P. Mesh-to-BIM: From segmented mesh elements to BIM model with limited parameters; 2018

[31] Shen YT, Lu PW. Development of Kinetic Facade Units with BIM-Based Active Control System for the Adaptive Building Energy Performance Service; 2016

[32] Ignatova E, Zotkin S, Zotkina I. The extraction and processing of BIM data.

IOP Conference Series: Materials Science and Engineering. IOP Publishing; 2018. p. 062033

[33] Bueno C, Pereira LM, Fabricio MM. Life cycle assessment and environmental-based choices at the early design stages: An application using building information modelling (in English). Architecture, Engineering & Design. 2018;**14**(5):332-346. DOI: 10.1080/17452007.2018.1458593

[34] Shahsavari F, Koosha R, Yan W. In: Schnabel MA, Fukuda T, Haeusler MH, editors. Uncertainty and Sensitivity Analysis Using Building Information Modeling. Vol. 1. The Association for Computer-Aided Architectural Design Research in Asia (CAADRIA); 2019. pp. 615-624. Available from: https://www2.scopus.com/inward/record.uri?eid=2-s2.0-85068415632&partnerID=40&md5=8896bc32678a7ca87796eaef94e99021; https://www2.scopus.com/inward/record.uri?eid=2-s2.0-85068415632&partnerID=40&md5=8896bc32678a7ca87796eaef94e99021

[35] Kensek KM. Integration of environmental sensors with BIM: Case studies using Arduino, Dynamo, and the Revit API (in English). Informes de la Construcción. 2014;**66**(536):e044. DOI: 10.3989/ic.13.151

[36] Kensek KM. Teaching visual scripting in BIM: A case study using a panel controlled by solar angles (in English). Journal of Green Building. 2018;**13**(1):115-137. DOI: 10.3992/1943-4618.13.1.113

[37] Kensek K. Visual programming for building information modeling: Energy and shading analysis case studies (in English). Journal of Green Building. 2015;**10**(4):28-43. DOI: 10.3992/jgb.10.4.28

[38] Lim S, Buswell RA, Le TT, Austin SA, Gibb AG, Thorpe T. Developments in construction-scale additive manufacturing processes. Automation in Construction. 2012;**21**:262-268

[39] Le TT, Austin SA, Lim S, Buswell RA, Gibb AG, Thorpe T. Mix design and fresh properties for high-performance printing concrete. Materials and Structures. 2012;**45**(8):1221-1232

[40] Hwang B-G, Zhao X, See YL, Zhong Y. Addressing risks in green retrofit projects: The case of Singapore. Project Management Journal. 2015;**46**(4):76-89

[41] Bock T. The future of construction automation: Technological disruption and the upcoming ubiquity of robotics. Automation in Construction. 2015;**59**:113-121

[42] Wu P, Xia B, Pienaar J, Zhao X. The past, present and future of carbon labelling for construction materials—A review. Building and Environment. 2014;**77**:160-168

[43] Hager I, Golonka A, Putanowicz R. 3D printing of buildings and building components as the future of sustainable construction? Procedia Engineering. 2016;**151**:292-299

[44] Tay YWD, Panda B, Paul SC, Noor Mohamed NA, Tan MJ, Leong KF. 3D printing trends in building and construction industry: A review. Virtual and Physical Prototyping. 2017;**12**(3):261-276

[45] Sakin M, Kiroglu YC. 3D printing of buildings: Construction of the sustainable houses of the future by BIM. Energy Procedia. 2017;**134**:702-711

[46] Jung Y, Joo M. Building information modelling (BIM) framework for practical implementation. Automation in Construction. 2011;**20**(2):126-133

[47] Gosselin C, Duballet R, Roux P, Gaudillière N, Dirrenberger J, Morel P.

Large-scale 3D printing of ultra-high performance concrete–a new processing route for architects and builders. Materials & Design. 2016;**100**:102-109

[48] Liu H, Holmwood B, Sydora C, Singh G, Al-Hussein M. Optimizing multi-wall panel configuration for panelized construction using BIM. In: Proceedings of the 2017 International Structural Engineering & Construction Conference (ISEC), Valencia, Spain; July 2017. pp. 24-29

[49] Azhar S. Building information modeling (BIM): Trends, benefits, risks, and challenges for the AEC industry. Leadership and Management in Engineering. 2011;**11**(3):241-252

[50] Bryde D, Broquetas M, Volm JM. The project benefits of building information modelling (BIM). International Journal of Project Management. 2013;**31**(7):971-980

[51] A. F. 360. Product Development has Changed. So Should the Tools. Available from: https://www.autodesk.com.au/products/fusion-360/overview [Accessed: 11 September 2019]

[52] Moledina MMG, Pin GW, Enegbuma WI, Ali KN, Adenuga K. Building information modelling technological innovations in industrialised building systems cost estimation. In: 2017 International Conference on Research and Innovation in Information Systems (ICRIIS). IEEE; 2017. pp. 1-6

[53] Autodesk. Export your Revit Model to a STL File. Available from: https://knowledge.autodesk.com/support/revit-products/learn-explore/caas/CloudHelp/cloudhelp/2020/ENU/Revit-AddIns/files/GUID-93AE8701-3958-43E9-8D95-0C1650B88061-htm.html [Accessed: 10 September 2019]

Chapter 9

An Investigation of Virtual Reality Technology Adoption in the Construction Industry

Mohsen Ghobadi and Samad M.E. Sepasgozar

Abstract

While Virtual Reality (VR) technology has experienced a recent growth in interest and offers immense potential in a number of domains, there is still insufficient information on the acceptance and adoption of this technology among individual users. The purpose of this chapter is to examine the acceptance and adaptation of people using VR technology in the construction industry and to identify factors that prevent VR technology from being adopted more widely in the construction industry. Semi-structured interviews were conducted to approach this research problem among 15 students and academic staff members at two universities. The results of this research indicate that VR technology is acceptable to people who work in the construction industry. However, there are barriers to further adoption of VR technology, namely high VR hardware and software requirements, low affordability, and low accessibility. This research also proposes several resolutions to these barriers, including preparing facilities by construction industries and universities, providing software and hardware requirements for VR technologies, and decreasing the price of VR devices. The results of this research are of immense value to suppliers and companies affiliated with this technology. Further research is required to demonstrate the functionality of VR technology in the construction industry.

Keywords: Virtual Reality, adoption, barriers, construction management, construction industry

1. Introduction

Virtual Reality (VR) technology allows users to interact with different objects in a virtual environment [1]. Shen et al. [2] found that virtual simulation, through VR technology, creates significant advantages including cost savings, time-savings, and improvements in training efficiency and safety in areas such as healthcare, construction, and manufacturing. Sherman and Craig [3] define VR technology as a medium composed of computer simulations that sense the user's actions and positions and change or augment the feedback to the user's senses, creating the feeling of being present in the simulation, in other words, being mentally immersed. Some VR applications aim to combine virtual displays with the physical world to create more meaningful virtual presentations by providing the ability to interact with real-life

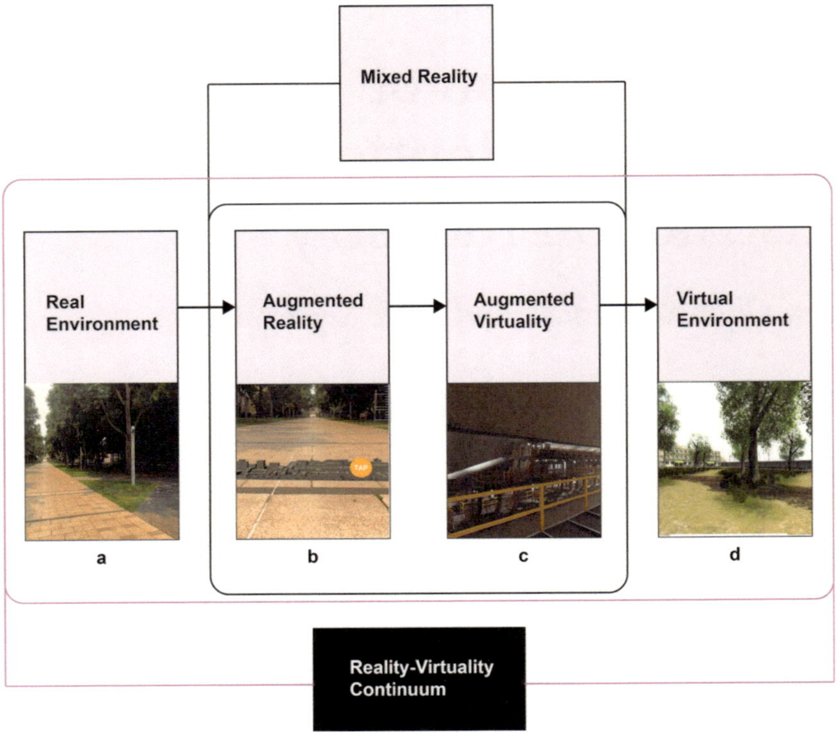

Figure 1.
Reality-Virtuality continuum that indicates the stages between real and virtual environment. (a): real environment (RE); (b) using mobile application to create Augmented Reality (AR) in real environment; (c) Augmented Virtuality (AV): virtual environment created from a real environment; (d): virtual environment (VE); Mixed Reality (MR): the combination of AR and AV.

stimuli and VR simulations. This is referred to as Augmented Reality (AR) and Mixed Reality (MR). AR allows the user to perceive the real world with overlaying virtual elements. MR presents the best qualities of AR scenario in an immersive interface to overlay upon reality [4]. **Figure 1** offers an illustration of the differences between VR, AR, and MR scenarios in the form of a Reality-Virtuality continuum.

The use of VR technologies has expanded in various industries in recent years [5–9]. One of these is the construction industry, although the adoption of VR technologies in construction has been limited. Some construction companies use AR to identify hazards and falling of their workers in their workplace [10, 11]. Based on the research of [10–13], there is significant potential for the wider adoption of VR and AR devices in the construction industry. VR may be helpful for training construction workers at low cost, with fewer safety risks, and faster methods in a virtual environment [12]. VR may also be a beneficial tool for architectures when building maquettes in the virtual environment, and may allow more creativity and time-savings [13]. Section 2.4 offers a detailed discussion of the application of VR and AR technology in the construction industries.

1.1 Outline of the chapter

This chapter provides information on the context of VR and AR technologies in the construction industry. This chapter aims to provide an in-depth understanding of the current and future opportunities of VR and AR technologies generally, and to examine the actual uses of VR technologies at present. The chapter is organized as follows:

- Problem statement, research goals, and research objectives will be discussed in detail.
- A review of the literature on VR technologies is presented, providing a summary of VR technology concepts as well as a review of the existing literature about the adoption of VR. The current potential barriers that affect the adoption of VR are also discussed.
- The research methodology is presented, involving semi-structured interview questions and a participant survey.

1.2 Problem statement

VR has existed for decades; however, only recently has VR technology attracted substantial and sustained interest in a number of domains. VR technology offers immense potential for various industries, such as manufacturing, construction, healthcare, education, and media [5–9]. Based on the research [14], only 7% of businesses currently use VR in their sector, but 23% of companies have a plan to use VR in the next 3 years due to an increase in consumer purchase intention and product value. **Figure 2**, taken from Google Trends, illustrates the recent trend of VR, AR, and MR technology interest in Australia. As **Figure 2** illustrates, the interest in VR, AR, and MR technology has been fluctuated in the past decade. In some years, visible in red, interest in these technologies has increased steeply. Interest in VR, AR, and MR technologies significantly increased in five different years, 2010, 2013, 2015, 2016, and 2017 [15, 16]. The reasons for some of these fluctuations are presented in **Figure 3**. For example, in 2010, AR technology was used for advertising purposes in the print media for the first time. This has

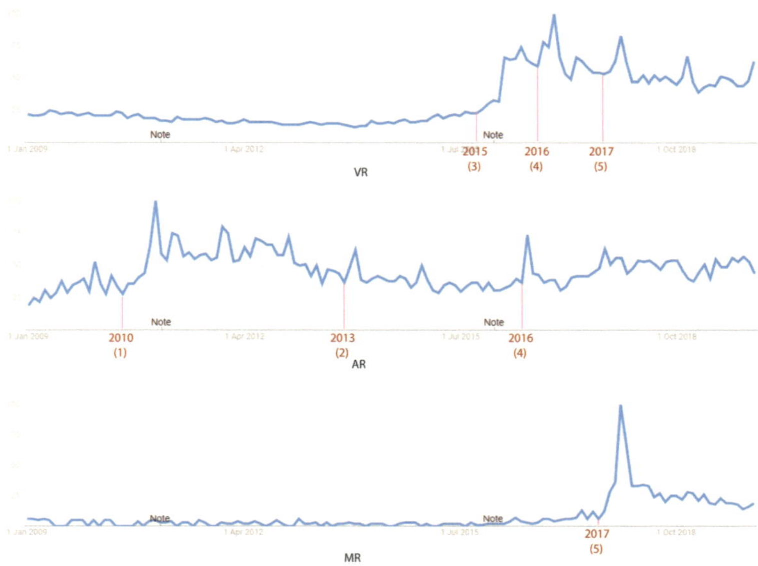

Figure 2.
VR/AR/MR interest trend since 2009 in Australia based on Google Trends. Note: the numbers one through five represent the years 2010, 2013, 2015, 2016, and 2017 with the growing interest of VR, AR, and MR.

dramatically increased the number of people who are familiar with VR technology [16]. In 2017, the first MR technology-related smartphone was introduced and released to the global market, the effect of that can be seen in the MR graph in **Figure 2** [4]. In addition, information obtained from CCS Insight about global shipments of VR and AR is displayed in **Figure 4**. As is illustrated in **Figure 4**, the global shipment of standalone VR and AR devices is on the rise. It is worth noting that the VR technology shift is higher than AR technology.

As is suggested by the figures above, awareness and engagement with VR technology are increasing, but this upward trend is not constant and is oscillatory. The reasons for the variegating interest in VR are unclear. Despite significant investments in the VR technology advancement, there still appears to be a reduced desire to use VR technologies across several domains [1]. While, various research projects [1, 17–19] have investigated the barriers to the adoption of VR technology in different industries separately, there are few resources to address the acceptability and adaptation of VR technology in the construction industry. Identifying the barriers and factors encouraging the adoption and acceptance of VR technologies in the construction industry is thus a vital issue for VR technology suppliers as well as potential VR user groups and VR-related businesses.

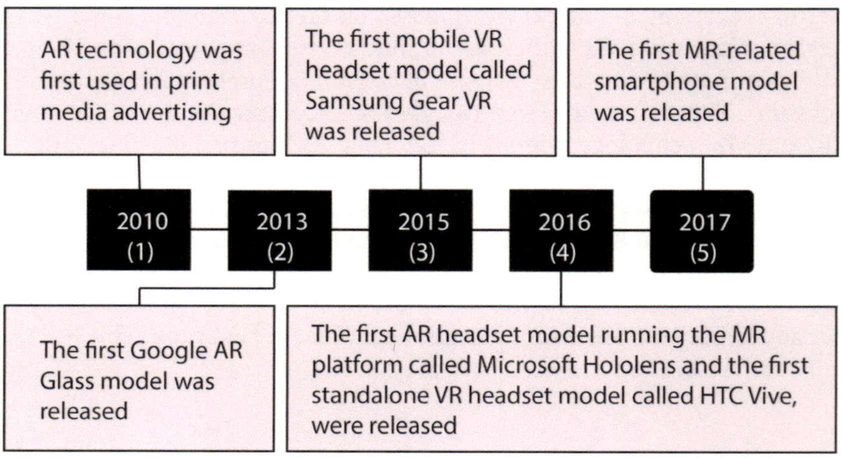

Figure 3.
Reasons for the growing interest of VR/AR/MR in five marked years illustrated in **Figure 2**.

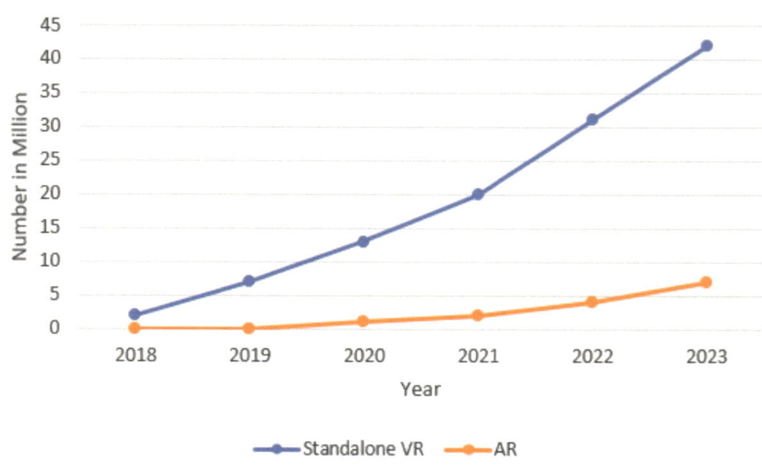

Figure 4.
Global shipments of VR/AR devices, data obtained from CCS Insight.

1.3 Research aims and objectives

There is no doubt that the interest in VR technologies is growing and VR technologies are being implemented in new ways across myriad domains. VR technologies are still in the early adopter stage and diverse applications for significant adoption are uncertain. Although VR technology can be a very applicable technology in the construction industry and there has been a lot of investment in this technology so far, VR technologies continue to be under-utilized in the industry. Some articles do identify some of the factors and barriers that affect VR adoption [1]. However, there are inadequate data to investigate VR adoption in the construction industry. Eliminating the barriers that hinder the growth of VR technology and developing the drivers for wider adoption are essential for wider application of VR technology in the construction industry. This chapter aims to examine the essential drivers and barriers to the adoption of VR technology in the construction industry.

The specific goals of this chapter are listed below:

- Indicate the current status of adoption and acceptance of VR and identify the opportunities of these technologies in the construction industry.
- Identify the barriers to the adoption of VR technologies.
- Identify the factors and drivers of adoption of VR technologies.
- Present solutions for eliminating VR adoption barriers.

2. Literature review

This section provides an overview of the literature discussing VR and AR technology, their applications in different areas, specifically in the construction industry. **Figure 5** illustrates the different VR platforms that we will discuss in the sections specified.

2.1 VR/AR technologies

The key terms and definitions related to VR technologies are presented in **Table 1**.

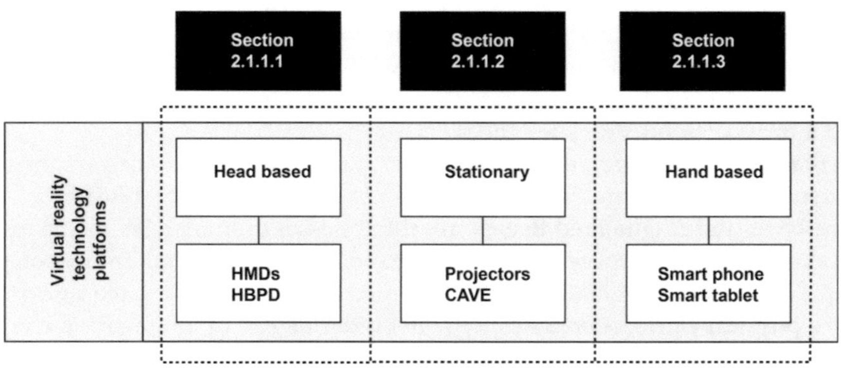

Figure 5.
Virtual Reality technology platforms. Note: HMDs: head-mounted displays; HBPD: head-based projector display; CAVE: the CAVE system.

Terms	Definitions
Virtual	Being or effect that is almost real, but not physically [20]
Virtual environment (VE)	A computer-generated environment within a simulation [21]
Reality	The state or quality of being real [3]
Immersion	The sensation of being present in an environment [21]
Virtual Reality (VR)	A computer-generated simulation of a realistic experience [22], where users interact practically and physically with the environment [23]. Also, a technology that produces the simulation via applying immersion theories into a virtual environment [24]
Augmented Reality (AR)	A similar concept to VR, where computer-generated content is added onto or embedded into the real-world experience so that both can be experienced simultaneously [22], a medium in which real-time interactive digital information is overlaid on the physical world that is in both spatial and temporal registration with the physical world [3]
Mixed Reality (MR) or (XR)	Incorporation of virtual computer graphics objects into a real-world three-dimensional scene, or the inclusion of real-world elements into a virtual environment [25]
Artificial Reality	Perceives a participant's action in terms of the body's relationship to a graphic world and makes responses that keep the illusion that his or her actions are happening within that world [3]
Mediated Reality	The concept of filtering reality, typically using a head-mounted video mixing display [26]
Cyberspace	A virtual network that exists to facilitate communication between users [27]
Telepresence	The ability to directly interact with a physical reality [27]

Table 1.
Key terms and definitions related to Virtual Reality.

Virtual Reality (VR) simulations are designed to create immersive worlds from which users have unique insights into how the real world works [27]. The critical elements in experiencing VR are participants, creators, the virtual scene, interactivity, and immersion [3]. First, participants are the most crucial foundation for experiencing this technology. The entire virtual scene happens in the minds of the participants and the experience in the virtual arena is tailored according to the participant's needs. The second most important element is the person who designs and implements VR technology as a creator. Third, immersion involves a feeling of being mentally or physically present in an environment. Mental immersion is the primary purpose of major media designers, while physical immersion is one of the characteristics of VR. Finally, interactivity is the response to a user action, which is essential for successful VR experiences [21].

Virtual Reality, Augmented Reality, Cyberspace, and Telepresence are the forms of computer-mediated interfaces of virtual and real worlds. Virtual Reality, Telepresence, and Augmented Reality are three classes of immersive media in a physical manner. Cyberspace, on the other hand, indicates mental immersion with other individuals. Augmented Reality (AR) mixes computer-generated information with the physical world. Mixed Reality (MR) is defined as an immersive technology that merges VR and AR [27]. Telepresence is defined as the ability to interact with a physical world without limitations on the size or position of the device utilized to transfer the user's command [27].

2.1.1 VR technology platforms

VR technology is a virtual scene that is generated by software in which the operator can feel and experience a virtual scene that reflects the real world. A head-mounted display (HMD) is a wearable device, which permits the user to explore the virtual area by moving one's head. The user can interact with individual controllers with the environment in many cases. Users can display video-recorded or computer-generated content with 360-degree cameras in VR. VR also can be categorized into three different groups based on their technology platforms: head-based, stationary, and hand-based [3]. Below, each of these platforms is described in detail.

2.1.1.1 Head-based

A head-based group such as helmet, HMDs, or HBPD does not allow users to see the outside view of the world. Users stand in front of the position tracking sensor and see the virtual world that is created by the computer system. The position tracking sensor recognizes where the user is looking and transfers that into the computer. Instead of wearing a headset, users can wear a projection system on their head and see the virtual world on the surfaces ahead of them. This is called a head-based projector display.

2.1.1.2 Stationary

A stationary group such as projectors and CAVE creates a location in space for users to feel the virtual environment without carrying any hardware. One of the most commonly used platforms is the CAVE system. CAVE system provides a space, such as a room that displays computer-generated imagery around it. Users are in the middle of this place and find themselves in a seemingly real environment.

2.1.1.3 Hand-based

Hand-based groups such as smartphones and tablets require users to hold and look at something like a mobile phone, monitor, or tablet in order to experience a virtual environment [3]. In some cases, headsets in the head-based group are capable of being mobile-connected. However, this type of technology restricts the user to using a mobile or tablet for being in a virtual environment.

2.2 Development of VR and AR

This section discusses the development of VR technology to highlight the crucial stages in creating the current generation of VR technology. **Figure 6** presents a continuum of the development and predicted future direction of VR technology [3, 15, 16, 28, 29].

As is illustrated in **Figure 6**, interest in and development of VR technologies has ebbed and flowed since initial experiments with reality-augmentation in the late 1950s. In the 1990s, when 3D gaming became more accessible and popular, there was a similar boom in interest in Virtual Reality. Gaming companies presented 3D videogames; movies similarly depicted virtual worlds and computer-generated environments. However, this technology is rudimentary and limited, with mediocre graphics, low computing power, time lags, and high prices. These products eventually failed because consumers were unhappy with these technologies and interest

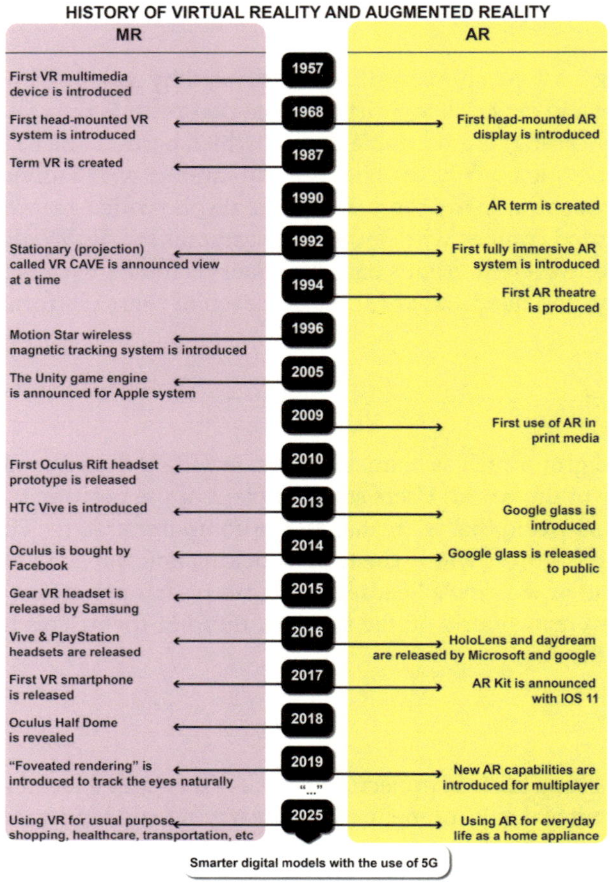

Figure 6.
Timeline of Virtual Reality and Augmented Reality technologies and the predictable future of them [3, 15, 16, 28, 29].

significantly diminished [30]. Since the 1990s surge in VR interest, new technologies have encouraged renewed interest and investment in VR platforms. Computers are powerful enough to make virtual worlds realistic, and mobile technologies have made higher quality VR simulations more available and accessible. New companies have become interested in this progress, and have invested significantly in new VR technologies [30]. For example, in 2014, Facebook made an investment of US$2 billion to buy the Oculus Rift, a VR headset from a technology-based company that combines a headphone with a microphone to help users feel more immersed in the virtual environment [31]. Since 2018, more than 225 VC investments in VR/AR have been made, raising $3.5bn inequity [30]. Flavián et al. [21] predicted that VR, AR, and MR sales in 2020 would grow to approximately US$ 61.3 billion, 21 times higher than VR, AR, and MR sales in 2016.

2.3 Potential sectors for VR/AR

There are various areas in which VR/AR technologies may deliver new potential to markets. VR technologies have sold well in the domain of personal entertainment, and statistics suggest that most investments have been made in these markets so far [30]. However, VR and AR technologies may offer further benefits in the architecture, engineering, and construction (AEC) industries. **Table 2** indicates the application of VR/AR technology in different sectors.

Sector	Application
Entertainment	Jung and tom Dieck [32] proposed a framework for visitors to use VR, AR, and 3D printing at a cultural heritage site to prepare a better experience for visitors. Mobile AR applications were used for tour guides. VR applications were used for simulating the inaccessible places. 3D printing was used for selling a small 3D souvenir. This offers cultural heritage sites new ways to attract visitors, encourage positive publicity, and enhance visitors' positive experiences, thereby encouraging future visits
Live events	Kim and Ko [5] conducted a case study to use VR as a media for spectators to see the sports match and make users feel like they are physically present at the event. By utilizing VR instead of 2D screen, the visitor's satisfaction increased. VR can increase the visitor's good experience by preparing more vitality, telepresence, and interactivity compared to the traditional method
Retail	Poushneh and Vasquez-Parraga [33] used an AR application for evaluating retail user experience by showing a different view of their designs to customers. AR can provide a new experience for customers to assist decision-making. This can affect customer satisfaction and their willingness to buy positively
Real estate	Ozacar et al. [34] created a VR real-estate application that delivers a computer-generated 3D simulation of realistic interior design of the on-sale unbuilt property. This VR application can satisfy the expectation of the property buyers or investors compared to 2D plans or renders
Healthcare	Hall et al. [6] evaluated the usability and acceptability of a 3D Virtual Reality technology to give health-related information to people with intellectual disabilities. This technology was used for a new way of healthcare-related learning. By utilizing a VR environment, people with intellectual disabilities could remember some features of that environment
Education	Chang et al. [7] indicated the use of AR in different fields of education such as chemistry, mathematics, spatial ability training, physics, geography, and surgical training in order to increase the level of quality and progress. AR technology can be useful by superimposing computer-generated virtual 3D objects over real environments
Engineering	Nee and Ong [8] found that AR has potential in different manufacturing activities such as robotics, product design, maintenance, CNC machining simulation, and facility layout planning. This technology offers high accuracy, desirable alignment with industrial standards, and fast response in the above-mentioned domains
Sport	Stinson and Bowman [35] utilized a VR system to present a virtual 3D scene for athletes training to feel a high-pressure training situation that can be helpful for athletes to become ready for a real job. This system could increase the anxiety in athletes' feeling compared to baseline condition and help them to be in a high-pressure situation like a real world
Architecture	VR can be useful in building and exploring 3D maquette models for architectures. It can help them in early design and ideation stages. Architects can create maquettes in a virtual environment by utilizing VR headsets. This can help designers to work more effectively, and at a faster pace compared to current CAD software [13]

Table 2.
Potential sectors for using VR and AR.

2.4 VR in the construction industry

VR and the development of virtual environments can shape the understanding of construction project stakeholders and their success in completing their projects. VR technologies can help improving the effectiveness and efficiency of detailed design, preparation and planning, and construction completion of a project. Increasing the understanding of the design intent, decreasing disruptions, and improving the constructability of the project can be achieved by reviewing and rehearsing the construction of the facility in a 3D immersive and interactive environment before the start of construction. **Table 3** indicates the application of VR technology in the construction industry.

2.5 Theoretical research framework

Several models have emerged that aim to explore the adoption and acceptance of new technologies. In **Table 4** these models are explained, and below, discussed in depth.

The theoretical framework used in this research considers the adoption of new technologies and employs the UTAUT model. This model assumes that performance expectancy, effort expectancy, social factors, facilitating conditions, and trust are the primary constructs in the adoption and acceptance of new technology among users. Performance expectancy is the most critical factor in utilizing modern technology. There are significant expectations for new technologies and uncovering the perceptions of users regarding the usefulness of VR technology is essential for meeting the demands of future users of VR technology. Effort expectancy considers the required effort in implementing and learning how to use new technologies. This construct can represent the ease of use in technology. Social factors consider

Section	Application
House building	Building a Virtual Reality system that allows architects to be immersed in a virtual environment for planning and building cities [36], investigating the use of VR technology for the simulation of on-site tasks in architectural practices [37], and utilizing VR for pre-sale housing system to decrease the project costs, quality risks, and delivery time. This system allows the user to experience the design of the unbuilt house [38]
Construction safety and training	Creating a Virtual Reality simulator of heavy mobile crane operations that can be employed for user training by doing the lifting process in the virtual world before the real project [12]; developing a framework to create training simulators of heavy construction machinery to improve productivity, safety, and quality perspective [9]; developing VR and MR learning and education program to evaluate the power of VR and MR to facilitate the knowledge acquisition [39]; developing a VR safety training program that is related to electrical hazards in the US construction environment [40]; and utilizing VR system to experiment the user's social behavior in the hazardous situation [41]
Project planning	Using a VR application in the construction job site for site layout planning, evaluation of construction site logistics planning, and collision detection [42], and utilizing VR and AR as a prototype tool to do the site planning, support construction layout, and rationalization of the logistics for increasing the productivity and decreasing the operating cost [43]
Augmented VR	Using a mobile AR to support monitoring and documentation of construction site improvement to visualize progress information [44]. Utilizing AR in the case of problem saving, design alternatives and decision-making through a design review activity [45]. Utilizing augmented 360-degree panoramas help users increase their hazard identification in the construction job sites [10]. Utilizing AR to deliver design layout to the electrical construction practitioners in a faster way with less mental and physical demand requirements [46], and utilizing the AR glasses for the users who are working in an assembly construction line [47]
Analyzing tool	Developing a framework to increase the efficiency of maintaining and developing gas and oil facilities by utilizing AR [48]

Table 3.
Using VR in the construction industry.

Model	Explanation
Technology Acceptance Model (TAM)	The TAM is mostly used in the field of Information Technology. The model assumes perceived ease of use and perceived usefulness are the most critical constructs influencing people's attitudes toward using new technology [49]
Unified Theory of Acceptance and Use of Technology (UTAUT)	The UTAUT model is one of the most widely used models for adopting new technologies. The model evaluates four primary constructs: (i) performance expectancy, (ii) social influence, (iii) effort expectancy, and (iv) facilitating conditions [50]
Real Estate Stakeholders Technology Acceptance Model (RESTAM)	The RESTAM is employed in the real estate online platform, considering 11 constructs: (i) information quality, (ii) service quality, (iii) system quality, (iv) playfulness and usability, (v) perceived enjoyment, (vi) perceived usefulness, (vii) self-efficacy, (viii) user satisfaction, (ix) perceived ease of use, (x) actual use, and (xi) behavioral intention to use [51]

Table 4.
The adoption of new technology.

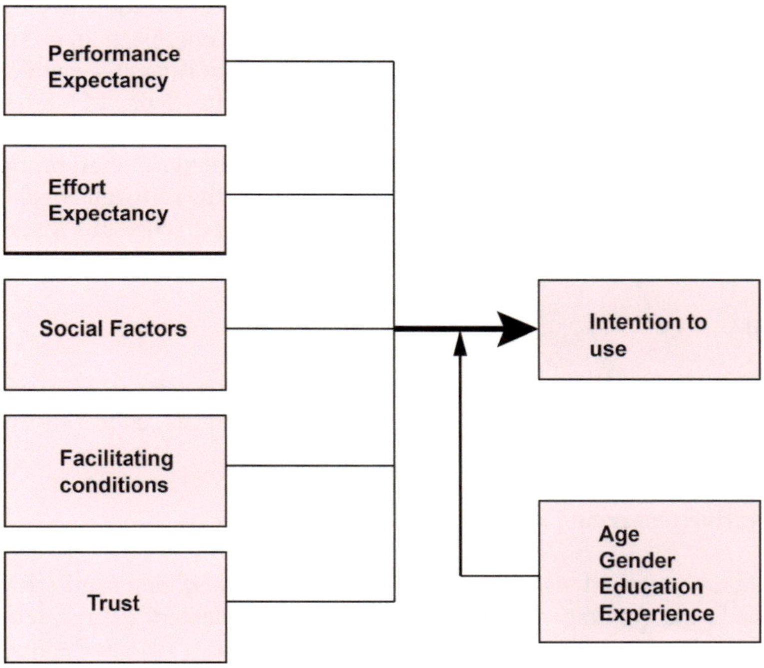

Figure 7.
Important constructs to the adoption of new technologies.

people's thoughts about adopting new technologies in other companies in the same industries. Facilitating conditions considers the functions of modern technology to simplify the existing conditions rather than complicate them. Trust involves identifying security and privacy risks and hazards of new technologies. The effects of these sub-constructs are evaluated during this research.

Figure 7 illustrates the relationships between the identified construct to the adoption of new technologies alongside other factors including age, gender, education, and experience. This figure is adapted from the UTAUT model [50].

2.6 Barriers to the adoption of VR technology

In this part, the obstacles to VR-based technologies are explained in light of the existing literature. Fernandes et al. [52] investigated the aspects that affect the adoption of VR among top UK construction firms in the public sector. The authors found the factors that affect VR by researching other resources; then, they organized these factors into various categories. The authors collected data from 33 top UK construction firms by distributing questionnaires. The results of this study indicate that the senior management support, champions of the technology within a company, degree of business competition, internal needs, the participation of users, and organizational resources are the most critical factors affecting the adoption of VR in the UK construction sector. Laurell et al. [1] identified the barriers to the adoption of Virtual Reality through social media. The author chose HTC Vive and Oculus Rift as empirical cases. The data were collected through all published user-generated posts in Swedish social media channels such as Facebook, Instagram, various blogs, and Twitter. The result of this study suggests that network, technology, price, and trialability are the most critical barriers to the adoption of Virtual Reality respectively. Glegg et al. [17], Paulo et al. [18], and tom Dieck and Jung [19] evaluated the effects of VR and AR adoption on the healthcare, tourism, and national parks sectors respectively. Stockinger [53] considered the current and future state of the VR and AR technology; however, the only participants of this study were VR experts. This sample may not be representative of the wider population, of industry actors and investors, with regard to knowledge of VR technology and its capabilities. Likewise, Whyte [54] only used experts as its participants to investigate users' contributions to the VR as new technology. Therefore, to fill the existing gap in the current research mentioned before, this research considers the perspectives of participants who have varying levels of expertise in VR technology.

3. Method

In the following section, the approach and reasoning for the selected method is outlined in more detail. The process of choosing contributors, collecting and analyzing data is also discussed.

3.1 Qualitative research

A qualitative method was selected for this research involving semi-structured interviews. In the qualitative method, information is collected, interpreted, and suggestions are made at the same time. Because individuals answer the questions, the themes that can arise may be dissimilar than expected. Qualitative approaches provide significant freedom for the examiner to highlight the outcomes of the study and the legitimacy of the gained data [55]. In order to realize and explore a broader phenomenon, qualitative methods are utilized by asking questions that allow participants to provide detailed and open responses. Developing and adopting new technology is one of the areas that needs qualitative research to gain better data from a broad selection of participants [55]. Ethics approval was sought and obtained from the Built Environment Faculty of the University of New South Wales.

Participants were asked to conduct the VR program, which was followed by a face-to-face interview. Participants used a VR program that was provided for the purposes of this study. The program was a Tunnel Boring Machine (TBM) simulator that allows users to see and interact with various locations and parts of the TBM. A laptop, an HTC Vive headset, and two handheld controllers were used to perform this simulation. Users can view the machine in real time through a headset, as well as being able to navigate through different parts of this machine via controllers. All the relevant information that the users needed to know to use the technology on their own was provided. **Figure 8** illustrates the HTC Vive headset that was used in this research. **Figure 9** shows a participant while taking part in the simulation exercise.

3.2 Case studies

Case studies were utilized in this study. This approach involved examining subjects and evaluating this new technology based on the identified construct such as performance expectancy, effort expectancy, facilitating conditions, trust, and social factors. The goal of the case studies-based approach is to examine a situation or an event from the perspective of the individual [56]. The case study approach aims to obtain different categories and themes, not to test hypotheses or prove relationships. In this sense, case studies involve an interpretive approach. Case study data were collected by doing interviews with participants [56].

3.3 Semi-structured interviews

Semi-structured interviews were selected in order to provide subjects with more freedom when answering questions and allow participants to provide information relevant to a number of VR applications and experiences. Completely structured interviews for this study could not be a useful approach because, as mentioned in

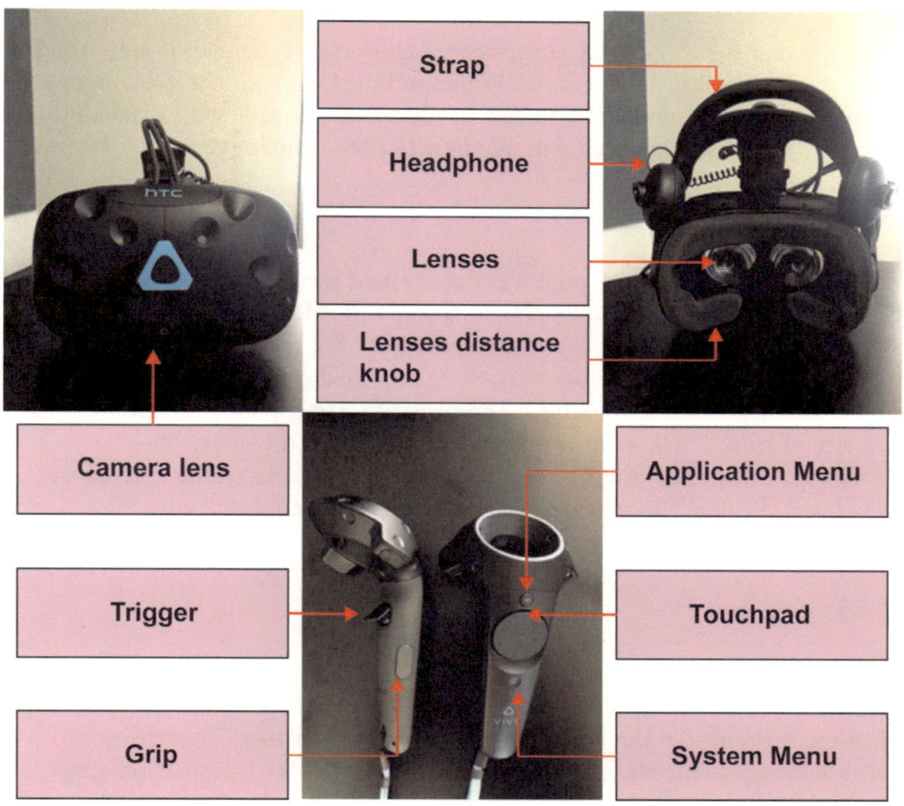

Figure 8.
Different components and parts of HTC Vive headset and controllers that are used in this research.

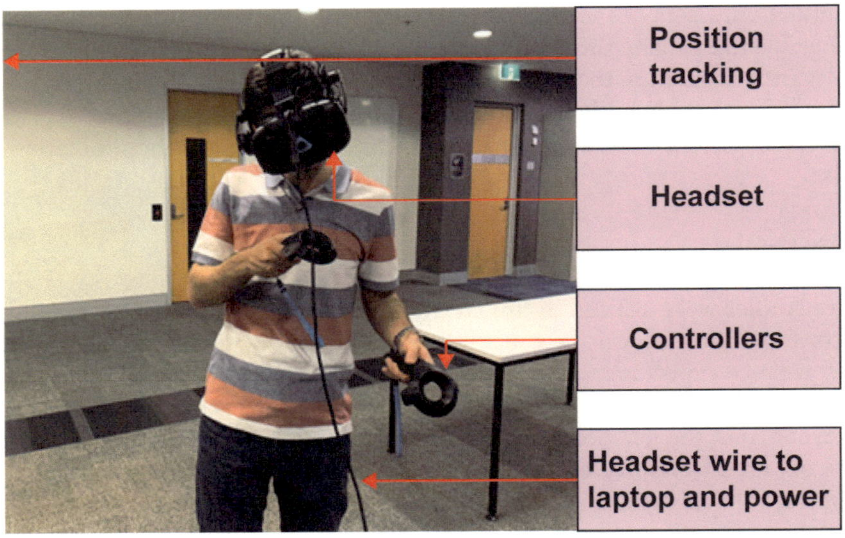

Figure 9.
Research participant experiencing Virtual Reality.

the literature review, there are many areas in which VR technology plays a role and the ways users experience and interact with VR technology are often difficult to predict.

3.4 Sampling

Interviewees for conducting this research chosen from academic staff and students at the University of New South Wales and the University of Wollongong from the built environment and construction faculty. Some interviewees used VR in their daily work, while others were relatively new to the technology.

3.4.1 Sample size

The size of the participant pool was determined by other similar research projects. Studies such as [54, 57], which are closely related to the subject of this research, used 11 and 10 participants, respectively. So, it can be inferred that this is the usual and ordinary number of participants used in such analyses. In this study, 15 participants were assessed. Out of the 15 participants, 4 were from the University of New South Wales and 10 were from the University of Wollongong, the academic staff member participant was from the University of New South Wales. The participants had work experience and education in construction.

3.5 Data collection

All participants engaged in this study with informed consent. Students who were interested in participating in the study filled out the consent form from the Built Environment Faculty of the University of New South Wales. To perform this research with academic staff, researchers emailed with the information about the research. Additional information and the consent form were sent to academic staff members on how to conduct the interview process, and the necessary coordination made to provide the appropriate time for doing the interview. All 15 face-to-face interviews were conducted among participants who were students and academic staff (14 students and 1 academic staff). Participants had varying levels of experience with VR technology. Interviews lasted about 20–30 min.

Semi-structured interviews were conducted as follows. Initially, participants were given basic information about the purpose and subject of this research. Second, participants shared their experience of using VR technology. The participants were then asked questions related to the adoption of VR technology. During the face-to-face interviews, the participants' voices were recorded as well as some notes taken. After each interview, all the participants' comments and ideas were written and saved as a Word document. All the audio recordings for this research and their transcripts were saved. After conducting all the interviews, these files were easily accessible for the analysis part. In the following section, choosing the appropriate software for qualitative research will be discussed.

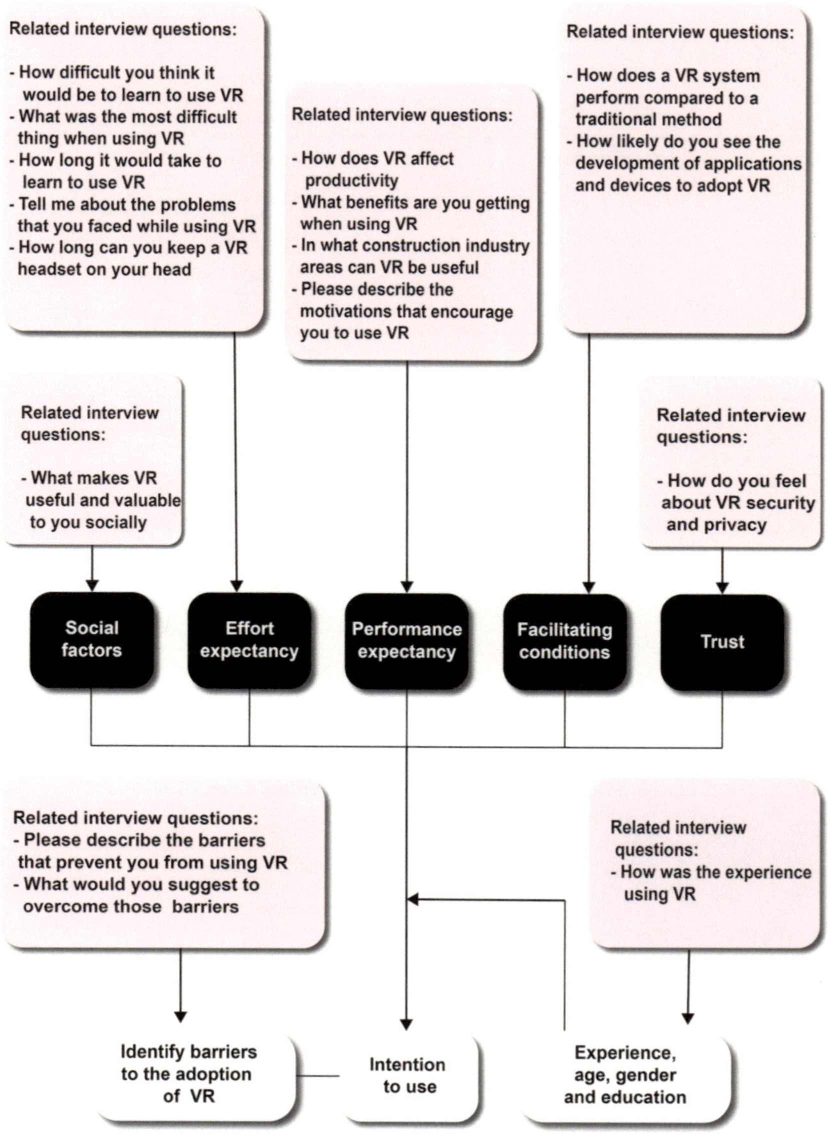

Figure 10.
The purpose of interview questions.

Figure 10 illustrates the purpose of interview questions that fit the needs of the survey to adopt new technology.

3.6 Data analysis

After all the interviews were completed, Word files provided after each interview were transferred to NVivo software. To perform the analyses, transcribed participant interviews were coded using the NVivo software. Thematic coding used in this research was based on the participants' responses. The final coding structure is available to see in **Table 5**.

ID	Factors	Codes
1	Application development	More effective
2	Difficulties	Depends on application
		Feel uncomfortable
		New technology confusing
		No difficulties
		No spatial awareness
3	Most difficulties	New technology
		No difficulties
		Setting up hardware and software
		Using headset and controller
		Working with some applications
4	Benefits	Be in a virtual place
		Ease of learning
		Involve different area
		Safety
		Time and cost saving
5	Problems	Application and program
		Feel uncomfortable
		Limited area of moving
		Setup correctly
		Using headset and controllers
		No problem
6	Usage area	Building construction
		Education
		Mining
		Site planning
		training
7	Barriers	Affordability
		Emotional and mental issue
		Hardware
		Not accessible
		Software
		No barriers

ID	Factors	Codes
8	Keep headset	Depends on the application
		Less than 30 min
		More than 30 min
9	Overcome barriers	Improve hardware
		Improve software
		Mix with traditional
		Reducing price
		University or industry facility
10	Drivers	Be in virtual space
		Different usage area
		Ease of learning
		New technology
		Not falling behind technology improvement
		Reliable
		Time and cost saving
11	Privacy	Depends on the application
		Privacy risk
		Same as other technologies
		No privacy issue
12	Socially	Communicate with others
		Depends on the application
		Flexible working conditions
		Share knowledge and information
		No benefits
13	Compared to traditional	More difficult
		More helpful
		More interactive and real
		Same
14	Experience	Better understand in learning
		Enjoyable
		High performance in future
		Interesting
		New experience
15	Productivity	Ease of learning
		Time and cost saving

Table 5.
Final coding structure.

4. Findings

After completing data collection and analysis, the results were divided into different categories according to the factors discussed in the research framework.

This section will explain participants' responses related to each factor as well as the barriers that prevent VR technology from being accepted by the participants.

4.1 Performance expectancy

This factor examines the usefulness of VR technology in the construction industry. To illustrate the value of this technology, participants answered four different questions including the effects of VR on productivity, benefits of using VR, different uses of VR, and drivers that encourage the use of VR. In the following section, participants' conversations regarding each question are reviewed.

4.1.1 Productivity

When participants were asked to comment on the effect of VR technology on productivity, all participants had a favorable view, saying that VR technology could significantly increase productivity, especially in the context of construction training. It is no longer necessary that actors, such as engineers, go to the original construction site to familiarize themselves with devices and machines. Below, one of these responses is provided:

> *"Education is where I think it does. Education and training. I think to try conveying the same thing through as I was saying, books, and lectures; you just cannot do it as quickly as you can with your hands-on things. The example is if Michael Jordan was doing a seminar on basketball, where no one played basketball" (participant number 4).*

Also, other participants felt that VR technology could impact productivity by time- and cost saving.

> *"In terms of productivity, I should mention that, for example, there is no need to be in the site because of course, it will be more cost extensive, and you can just use it in a room and you will see the Industrial site and whatever and it will save more time, cost and of course it will lead to higher productivity" (participant number 14).*

So generally, they thought that VR technology could have an impact on reducing costs and time as well as enhancing the ease of learning for construction engineers.

4.1.2 Benefits of using VR

When participants were asked to comment about the benefits that they were getting while using VR, most of them mentioned that it was an excellent opportunity for them to be in a virtual world, which offered opportunities related to time and cost savings and safety. Below, one participant's response regarding this question is provided:

> *"Well, of course, you can be somewhere in a virtual world which can be difficult to be there in reality or dangerous or even impossible because of restrictions. And that you can be anywhere you need to be with no matter of cost and time" (participant number 15).*

Participants also postulated that VR could help them to have a better understanding of a different area of knowledge.

> *"Well, I think VR technology involves many other technologies in different disciplines like computer science and or, construction management or other knowledge, so it has encouraged me to learn more knowledge in terms of other disciplines. So, I think this is a benefit I am getting" (participant number 3).*

4.1.3 Different usage areas of VR

Once participants were asked to think about an area of the construction industry in which VR can be helpful, all identified construction training and education as the first area. Below, we look at one participant's response regarding this question.

> *"One of the main ones that as a mining or civil engineer working in underground, or excavation like mining or tunnelling, which is not easy to get there by engineers. It means you cannot even get there. So, it helps engineers. Engineers in this field or technicians to get a better understanding of what is happening underground. Because We have a lot of problems when we go underground and when we want to start working there, so we must just do a lot of introduction, a lot of inductions to engineers for the first time. So, it gives us gave them a chance just prepared for the dangers" (participant number 10).*

Some participants also named other areas of the construction industry such as site planning, mining, building construction could be affected by VR technology.

> *"Sometimes, for example, for some construction site [s] that still on the progress on their construction that they are not ready but we want to present it for the clients for the stakeholders it would be interesting and also attractive, it will make them satisfied to you know, have a look at the future construction port construction site, how we will be the shape how are the features and these things that are the most part used" (participant number 14).*

4.1.4 Drivers encouraging the use of VR

When participants were asked to provide ideas about the drivers that would encourage them to use VR technology, most participants mentioned that being immersed in a virtual space where they cannot be in reality is the key driver for VR adoption. Some participants thought that VR is a new technology and they like to use that and if they do not use that they lag behind technology advances:

> *"I think in the last few years, it is the same as that. If you do not get into VR, you are going to fall behind businesses and the research and other areas like that. When you need big emotional things like emotional experiences or complex learning and teaching kind of situations where movement is important to you, I think then it is good. But I do see that in a lot of cases, AR will probably be more useful" (participant number 4).*

Furthermore, some participants mentioned another time that VR eased learning, that VR technology has different usage areas, and it offers time and cost saving and reliability. Hence, these are the essential drivers that encourage the use of VR in the construction industry.

4.2 Effort expectancy

This factor examines the level of ease of using this technology in the construction industry. Participants answered five different questions including identifying the difficulties of learning to use VR, time spent on learning, problems arising during use, and the amount of time that they can hold the headset. In the following section, participants' conversations about each question reviewed.

4.2.1 Difficulties of learning to use VR

Once participants were asked to share their ideas about the number of difficulties that they encountered to learn to use VR technology, almost half mentioned that there were no difficulties to learn how to use VR technologies. One answer regarding this statement is as follows:

> *"I think for people who know about basic thing about VR they can do it and no need to be deep and I think it is easy to learn" (participant number 11).*

Some participants thought that learning to use VR is relatively easy, but depends on the sort of applications.

> *"Not difficult at all. It was fundamental; it depends on the application itself. But for example, from what I saw, it had like few only like, buttons and few others like, stuff to do. It was straightforward. It was like; you just must try and error. You can learn a lot about to know how to use the whole application. So, it depends on the application" (participant number 13).*

Other participants felt confused because VR technology is a new technology that they had not used. Other participants felt uncomfortable because they have difficulties with spatial awareness. Participant number 4, who is a member of academic staff, offered some insight into this difficulty:

> *"There is a flat statistic of about 10% of the population who just will never feel comfortable. It is something to do with the physical. Functionality arises or something like that. So that flat statistic already says 10% is going just never to get a no matter how long put it on."*

We can conclude that some people had difficulties with using this technology that could be solved over time, while others will continue to feel uncomfortable.

4.2.2 Most important difficulties with learning to use VR

Participants were asked to give their opinion on the most crucial difficulty that they experienced when using VR technology. Some participants had problems with handling the controllers. Others felt that headset was heavy to use, because of the kind of headset that was used in this research. Again, some participants felt that VR is a new technology. Below, two participants' responses regarding this question are provided:

> *"Because VR is like a new technology for us as both students like our generation like most of the time that we are using VR so might be like having different feelings about the headset also the controller because I have never used the controller before. It is like our real life is still take some time to learn it was quite simple like I can actually learn in five minutes and pretty straightforward" (participant number 2).*

> *"Well, I think the most difficult things might be I think it might be people. Some people might be reluctant to learn new technology in they do not want to try new things. Because they feel unsafe or hard to learn new technology. So, I think this might be one of the difficulties for learning VR" (participant number 3).*

So generally, some difficulties in using VR were observed, but it is worth pointing out that these issues were considered to be easily overcome with further training and exposure to VR technology.

4.2.3 Time spent learning

Participants were asked to indicate how long it took to learn to use VR technology alone. Almost all participants acknowledged that it only took a few minutes to get to know everything. One answer from participants is as follows:

> *"It is a couple of seconds it depends on how you adapt yourself and how fast you are" (participant number 12).*

4.2.4 Problems arising during use

Some participants mentioned that problems using VR depend on the application that they need to use, because, if it is complex, using controllers will be difficult. Correctly setting up the system may be another issue checked before starting the application. One aspect that some participants mentioned about using VR was the limited area of moving. We can see one participant's response to this question.

> *"For me, as you saw, I have a problem that I cannot control my body. When I am using a headset, and I need to be restricted to an area just to prevent any kind of hurt because I might hurt myself and I am going to be using VR. I have no idea where I am. So, I start moving around" (participant number 10).*

Other participants mentioned that they did not encounter any problems while using this technology. Essentially, some problems that participants faced when using VR technologies require further research.

4.2.5 The amount of time the headset holds

Interviews found that almost half of participants were willing to keep the headset on for an extended period, but others believed it was difficult for them or the type of program they were using should be so attractive that it makes them eager to keep that. One sample answer is provided below:

> *"The one that I tried with you brought is for us, that was pretty and good the headset was comfortable but from the point that we are human, I think for me, maybe 25 to 30 minutes. I am going to need to feel safe, I mean, the real reward but I think that can gradually increase the thing but if the headset we put them be so much*

uncomfortable. That is going to be annoying or make you headache or earache" (participant number 6).

4.3 Social factors

This factor examines the participants' thinking about adopting VR technology in other companies in the construction industry. Participants were asked about their ideas regarding the issues that make VR useful and valuable to them socially. Most participants thought that other businesses in the construction industries could also use this technology and be at the same virtual place, it would be important and valuable for them to have communication with each other during experiencing a scenario. Participants also thought that it would be useful for sharing and transferring their knowledge and ideas in a flexible working condition. Two sample answers are provided below:

"One of the most important things is that how can we contact other people if we can make a connection, contact with other peoples around the world. So, it is going to make it so reliable and easier to get a lot of information and getting knowledge from all around the world, for example about one specific part. So, it is going to make it more enjoyable and more reliable. If we have other people involved in one field of interest" (participant number 9).

"It can be beneficial because I am sure that it can improve efficiency. But if we do not know the user, I am not sure how it can use if you want to know the idea of different people. It is complex or risky. You are not sure who they are. It depends on the goals and what you want to do I am sure that some people use that for online gaming maybe it can be good" (participant number 7).

4.4 Facilitating conditions

This factor examines whether VR technology facilitates the existing methods such as those used in education and training. To illustrate this situation, participants were asked to share their ideas about comparing VR technology with traditional methods and developing VR applications and devices.

4.4.1 Comparing VR technology with the traditional method

Once participants were asked to talk about their ideas about comparing VR technology with the traditional method, most compared using the headset and controllers with using the mouse and keyboard. However, some participants compared VR technology with lectures and books in educational and training purposes. One sample answer is provided below:

"Honestly, I do not think there is a huge difference, where it comes in is the kind of spatial awareness thing. So, having a 360 view on a computer and YouTube, for example, is pretty much the same thing as a headset. However, the level of immersion goes up when you have everything down, and it is right up against your face. But I would say it is still similar that there is not that much difference at the moment between the situations where it is probably good to have a mention that I think is in training things, for example, I think that is when it helps me. It is hard to translate tacit knowledge. And I think VR is a good way of allowing you to have the experience, without that, you cannot read a book and learn a lot of things. Yeah, that is the kind of fills that gap" (participant number 4).

Some other participants felt that using VR headset and controllers can be more useful. One participant's response is provided below:

> *"I think when you use the VR technology, you feel better that you feel that you are literally in that situation or in that place. So, it helps you to understand the environment. Like you are there. But when you use the keyboard, and you just follow it on screen, yeah, it does not give the same feeling and it helps you to remember what you see in a better way because you feel it" (participant number 8).*

All participants thought that using VR technology could be more effective than previous methods.

4.4.2 Development of VR applications and devices

All the participants agreed that developing VR applications and tools can improve the adoption and acceptability of VR technology because this can help VR to promote requirements more than before. One sample response is provided below:

> *"I think it is a part of our future, you need to accept it. I think within the next five or ten years, it is going to be mandatory for each engineer to know how to use this If they want to find a proper job. It is much less money and time-consuming to train people in this way, rather than to take them to real facilities. And it is no experience-wise it is. It is much more useful. It is much more fun to be in a three-dimension environment rather than just sitting at a desk and using keyboards and mouse they need to look if you want to encourage someone to use it. The first thing you need to do is to make them use it for five minutes. Because once they experience it for the first time, they realize that, there is a significant difference between, sitting at a desk and using desktop and keyboards and using a VR. And because we are visual people, the whole visual experience is fundamentally different and better when you are using VR. So, there is a huge room for improvements" (participant number 5).*

4.5 Trust

This factor examines the security and privacy hazards of new technology. For analyzing this factor of VR technology, participants were asked to talk about the privacy and security of VR. Almost half of the participants mentioned that the protection and confidentiality of VR are the same as other technologies. One sample answer is provided below:

> *"I think this technology is just like the rest of the software that depends on the internet and doesn't make much difference so overall I think it's a safe technology" (participant number 15).*

Some other participants thought that this technology could be problematic regarding safety and privacy, but it depends on the applications. One sample answer is as follows:

> *"I think when you are playing VR, they are somewhere to stand near you because you cannot see the surroundings, and you might have some safety issues. For our privacy issue depends on the software developer" (participant number 1).*

4.6 Experience, gender, age, and education

There are several sub-factors such as experience, gender, age, and education that need consideration to give a precise and accurate view of the adoption of VR technology. After the interviews, it appears that the gender and education factor did not affect any of the above situations, but the age and experience factors may be relevant.

Regarding the age factor, it is worth noting that participants believed that some people, especially older people, were not willing to learn new technologies and certainly learning new technologies can be harder for them. One sample answer is as follows:

> *"Elderly people are going to have many more problems with using that. It is going to be a total difference, you know, experience with them. To be honest, I cannot suggest anything, when you are dealing with older adults. One way or another they need to cope with it. It is a part of living in 2020" (participant number 5).*

Regarding the experience factor, participants who did not have previous VR experience found this technology to be a new, interesting, and enjoyable experience. Still, those who had more experience had a closer look at this technology and thought that this technology is a better way of learning and this will be used more in the future. Two sample answers are as follows:

> *"It was very interesting because I never used VR in that way. We use it for games, but I never used it with work industry related stuff. So, it was very interesting, and I thought like, you could not use it a lot in this field. But now I think there is a lot of things that you can learn from that" (participant number 13).*

> *"It was good. I was somehow familiar with the whole concept of using VR. But the thing was that it was my first time to use it for the specific purpose that if you are using it, it was interesting because I never had any kind of experience of being in that specific situation. And I enjoyed it. It did not last too much. It was something like 10 minutes" (participant number 5).*

4.7 Barriers to the adoption of VR

Participants were asked to express their views on barriers that impede the adoption of VR technology in the construction industries.

Several participants identified a lack of access to this technology as one of the critical barriers. They believed that this technology was not available and accessible to everyone, because of the high cost of the device and because this technology relies on a high-powered computer system that may not be widely available. One sample answer is as follows:

> *"Might be the equipment because like this kind of equipment is not common. You know for your own house. Also, like the computer that we need. Like it might be the computer with a better processor. Like we should not run just the big applications. You need a more powerful computer which is not accessible for some of the students" (participant number 2).*

Some other participants thought that the emotional and mental issues of some people who cannot use headset are the main barriers to the adoption of this technology. One sample response is as follows:

> *"Some of the population feel uncomfortable while wearing VR headset. It is related to the emotional and mental issues of them which is not about the duration of using that" (participant number 4).*

4.8 Overcoming current barriers to adoption of VR technology

One of the goals of this research was to recommend solutions to overcome the barriers to adopt VR technology. To accomplish this goal, participants were asked to suggest a way of removing and overcoming the barriers that they mentioned in the previous section. All suggestions made by these participants will be explained in this section.

4.8.1 Improve hardware

One of the factors that were considered by the participants as a barrier to the adoption of VR technology is the hardware used in this technology. It may be worth mentioning here that we did not use the latest headset and controller models in the market; instead, we used the available models, which had high power. It was concluded that over time the hardware available for this technology would improve and this can help a lot in adopting VR technology in the construction industry.

4.8.2 Improve software

Another factor in preventing the adoption of VR technology is the high-powered software needed to use such technology as computers with powerful resolution and the frame rate. The computer needs strong graphic processing unit power to be able to handle the VR applications. Indeed, one of the barriers that seem to be hard to fix is this factor because such computers are not available to everyone and suppliers of this technology need to find a solution to this obstacle to make this technology more acceptable in the construction industries.

4.8.3 Mix with traditional

One of the barriers raised in the previous section was the mental and emotional issues that some people have regarding new technologies. One solution that can help people use this technology is to mix VR technology with already established systems. Using the keyboard and mouse instead of using the headset and controllers is an example of that. This solution may help people to use this technology without any uncomfortable circumstances and it can also help older people to adopt technology more efficiently.

4.8.4 Reducing price

A critical issue that prevents the adoption of this technology among different people is the high costs associated with VR technology. Over time, it seems that there is a declining trend in the price of this technology. As such, this barrier is likely to be overcome naturally as VR technologies are produced more cheaply.

4.8.5 University and industry facilities

One of the solutions suggested by the participants of this research can help a lot in adopting and familiarizing different people with VR technology. This suggestion is that universities and construction industry centers provide this technology for use among their students and engineers respectively. This may solve both the problem of not having access to technology and the fact that people do not have to pay a lot of money to buy this technology and the associated software.

4.9 Summary

In this section, the responses from different participants in the context of the factors of the study were evaluated. **Table 6** offers a summary of the findings for each interview question. Fifteen questions were asked of participants of the study, and according to the participant's responses to each question, several codes were assigned to answers. The percentage column represents the percentage of participants who responded to the relevant issues by referring to each code. For instance, all participants agreed on the first question that VR technology can be instrumental in the construction industry. In answering question 6, 36% of the participants believed that VR technology could be beneficial and useful in the construction education section and 28%, 21%, 11%, and 4% of the participants, respectively, have referred to building construction, training, mining, and site planning as proper sections for application of this technology. An example column on the far right of the table provides an example of the participants' response to each question and code.

ID	Codes	Percentage	Example
1	More effective	100	"... It can be more effective in different areas ... "
2	Depends on application	16	"... It depends on the application itself ... "
	Feel uncomfortable	5	"... Felt uncomfortable when wearing a headset ... "
	New technology confusing	21	"... Felt confusing because VR technology is the new technology ... "
	No difficulties	53	"... No need to be deep and I think it is easy to learn ... "
	No spatial awareness	5	"... Felt uncomfortable because I do not have any spatial awareness ... "
3	New technology	25	"... Some people might reluctant to learn new technology ... "
	No difficulties	12	"... I can learn in five minutes and straightforward ... "
	Setting up hardware and software	6	"... Setting up software and hardware is the important part ... "
	Using headset and controller	38	"... Having different feelings about the headset also the controller ... "
	Working with some applications	19	"... It depends on the sort of application ... "
4	Be in a virtual place	30	"... Somewhere in a virtual world which can be difficult to be there in reality ... "
	Ease of learning	26	"... It has encouraged me to learn ... "
	Involve different area	9	"... Involves many other technologies in different disciplines ... "
	Safety	13	"... Somewhere in a virtual world which can be difficult to be there in reality or dangerous ... "
	Time and cost saving	22	"... Be anywhere you need to be with no matter of cost and time ... "

ID	Codes	Percentage	Example
5	Application and program	25	"… It depends on the application that we need to use … "
	Feel uncomfortable	20	"… I have no idea where I am … "
	Limited area of moving	15	"… I have a problem that I cannot control my body … "
	Setup correctly	5	"… Correctly setting up the system can be another issue … "
	Using headset and controllers	30	"… Using controllers will be difficult … "
	No problem	5	"… I did not feel any problem … "
6	Building construction	28	"… For some construction site that still on the progress … "
	Education	36	"… Engineers in this field or technicians to get a better understanding … "
	Mining	11	"… As a mining or civil engineer working in underground … "
	Site planning	4	"… Construction port construction site … "
	Training	21	"… It gives us gave them a chance just to prepare for the dangers … "
7	Affordability	13	"… Because of the high cost of the device … "
	Emotional and mental issue	9	"… Some people who cannot use headset … "
	Hardware	17	"… You need a more powerful computer … "
	Not accessible	9	"… lack of access to this technology … "
	Software	17	"… Need a better processor … "
	No barriers	35	"… There are no barriers … "
8	Depends on the application	18	"… It depends on how you adapt … "
	Less than 30 min	29	"… It is a couple of seconds … "
	More than 30 min	53	"… I did not feel any inconvenience … "
9	Improve hardware	12	"… There is a room for much improvement … "
	Improve software	39	"… Having computers with powerful processors … "
	Mix with traditional	12	"… Mix it with its previous versions … "
	Reducing price	12	"… Seeing a declining trend in the price … "
	University or industry facility	25	"… Universities and construction industry centres provide this technology … "
10	Be in virtual space	31	"… Be in a place that we cannot be there … "
	Different usage area	17	"… I do see that in a lot of cases … "
	Ease of learning	9	"… VR ease the way of learning … "
	New technology	17	"… Experiencing new technology … "
	Not falling behind the technology improvement	9	"… You are going to fall behind businesses … "
	Reliable	4	"… It brings reliability … "
	Time and cost saving	13	"… It brings time and cost-saving … "

ID	Codes	Percentage	Example
11	Depends on the application	18	"... It depends on the software developer ... "
	Privacy risk	12	"... You might have some safety issues ... "
	Same as other technologies	23	"... This technology is just like the rest of the software ... "
	No privacy issue	47	"... I think it's a safe technology ... "
12	Communicate with others	33	"... Contact with other peoples around the world ... "
	Depends on the application	6	"... It depends on the goals and what you want to do ... "
	Flexible working conditions	6	"... It is going to make it more enjoyable and more reliable ... "
	Share knowledge and information	11	"... Make it so reliable and easier to get a lot of information and getting knowledge ... "
	No benefits	44	"... It is complex or risky ... "
13	More difficult	6	"... It could be more difficult than traditional ... "
	More helpful	19	"... You cannot read a book and learn a lot of things ... "
	More interactive and real	69	"... You feel that you are literally in that situation or in that place ... "
	Same	6	"... I do not think there is a huge difference ... "
14	Better understand in learning	29	"... There is a lot of things that you can learn from that ... "
	Enjoyable	12	"... It was delightful because I never used VR in that way ... "
	High performance in future	4	"... It will be more useful in the future ... "
	Interesting	38	"... it was interesting because I never had any kind of experience ... "
	New experience	17	"... learning new technologies can be harder for elderlies ... "
15	Ease of learning	65	"... As quick as you can with your hands-on things ... "
	Time and cost saving	35	"... It will save more time, cost and of course it will lead to higher productivity ... "

Table 6.
Summary of the findings.

5. Discussion

The purpose of this research was to demonstrate the acceptance and adaptation of VR technology in the construction industry. To achieve the research goal, it was essential to examine VR technology based on different factors provided from the theoretical research framework. The research identified several barriers that hinder the adoption of VR technology in this sector. Various factors that drive different people to adopt and use this technology were also discussed. Finally, this research proposed various suggestions for removing barriers to further adoption of VR technologies in the construction industry.

Performance expectancy, effort expectancy, social factors, facilitating conditions, and trust were identified as the factors that determined the acceptability status of VR technology based on the theoretical research framework. Performance expectancy examines the usefulness of VR technology in the construction industry. By analyzing this factor, the results indicate the usefulness of VR technology in the construction industry. VR technology offers immense potentials for productivity. This technology is applicable in different parts of the industry and these features encourage people to use VR. Effort expectancy examines the ease of using this technology in the construction industry. In the effort expectancy factor, the results illustrate the high level of ease of using VR technology in the construction industry; there are some barriers to this factor. From the participants' point of view, it seems that the degree of difficulty of using this technology is not great, but due to some factors such as anxieties related to using new and unfamiliar technologies, it can be concluded that the required effort expectancy factor is not fully met by VR technology. Social factors include the participants' opinions about adopting VR technology in other companies. Based on the participants' responses, VR technology worked well on social factors in the construction industry and most of the participants feel excited that they can have a connection with others in the virtual environment. Facilitating conditions consider the ability of VR technology to be adopted into the construction industry. The results show that a high percent of participants agreed on the good performance of VR technology in facilitating conditions. Finally, the trust factor examines the security and privacy hazards of VR technology in the construction industry. Most of the participants agreed that VR technology could be like other technologies and there is not any hazard regarding the privacy issue. Still, there are some obstacles related to this factor to the performance of VR in the construction industry and some participants mentioned that the level of privacy hazard is related to the sort of the application.

There are some sub-factors that may also affect the adoption of new technology. These were age, experience, education, and gender. The results indicate that the gender and education factors were not affecting the adoption of VR technology in the construction industry, but age and experience factors can influence the adoption of this technology, because the level of experience makes people more focused on the technical issues of VR technology, which can affect the adoption of VR technology. The age factor can influence the learning and using of VR technology. A clear example of this factor is related to the elderly people. Some are not interested in using or learning new technologies.

The results of this study indicate that there are some barriers to adopting VR technology in the construction industry. These barriers include affordability, emotional and mental issues, hardware and software issues, and accessibility. Affordability relates to the high price of this technology. Emotional and mental issues refer to people's inability to use VR technology. Hardware issues relate to the computer requirements for using this technology. Software issues were related to the required high level of improvement in the VR technology and simulation applications. Accessibility of this technology suggests that VR technology is not available for all individuals. Each of the above factors needs to be addressed by VR technology-related companies and industries.

After identifying factors that prevent people from adopting VR technology, the participants also made suggestions for overcoming these barriers. These suggestions are as follows: improving VR hardware and software in construction sector; mixing VR technology with traditional learning methods to help some people who cannot shift suddenly from traditional methods to VR technologies; reducing the price of VR-related devices; and preparing VR technology facilities for engineers and students by construction industry companies and universities respectively. By addressing these issues, the process of adopting VR technology in the construction industry may accelerate, and VR technologies may become more common in the construction industry.

6. Conclusion

Based on the Findings section, the performance expectancy factor was analyzed based on the participants' responses for VR technology, which demonstrates the efficiency and usefulness of this technology in the construction industry. The effort expectancy factor was analyzed based on the participants' responses, demonstrating that the VR technology is relatively easy to use. The social element was analyzed based on the participants' responses related to the construction industry, which indicates the usefulness and effectiveness of this technology. The facilitating condition factor was analyzed based on participants' responses, which illustrates the high efficiency of VR technology in facilitating conditions. Generally, it seems that most of the participants felt that there is no serious threat to the security and privacy of VR technology. Based on the results on age, experience, gender, and education factors, it seems that gender and education factors do not have a significant impact on people's intentions of using VR technology. Still, age and experience factors do have some influence. Affordability, hardware, software, accessibility, and mental issue are some other factors identified as preventing the wider adoption of VR technology. However, improvements in hardware and software, reductions in price, and more availability of university and industry VR facilities may remove the barriers.

6.1 Limitations

There were some limitations to this study, including the following: first, due to the limited time available for conducting this study, few interviews were conducted for this study while many more interviews need to be undertaken to produce statistically meaningful results. Although other similar studies have had approximately the same number of participants, the accuracy of the results increases as the number of participants increases. Secondly, this research was conducted at only two universities in Sydney, Australia: the University of New South Wales and the University of Wollongong. This represents a relatively limited and narrow pool of potential participants, who may have greater exposure to VR technologies due to their socio-economic status, the wider availability of VR at Australian universities relative to the rest of the world, and a generally younger and more technologically savvy cohort. Third, before conducting all interviews, participants used one sample of VR technology, which indicated that they used only one type of application and program. The results may be different if participants were using other programs. Various programs have their target audience. Some programs may be less appealing than others; participants should be familiar with different types of programs so that they can give more accurate comments. Finally, the latest version of the VR headset and controllers was not used in this study, which may affect the results. With the advancement of VR technology, there have been changes in the applications, headsets, and controllers of VR technology since this study was undertaken.

6.2 Recommendation

Given the current state of VR technology and its associated hurdles, there is still much to be done on how to develop this technology. In the case of specific headsets currently dominating the market, different types of headsets can have different effects on research results. Further research is needed to reach more participants and use the latest version of this VR headset and controllers for future research. This study also identifies a more general need to examine the functionality of

Virtual Reality technology in the construction industry and to demonstrate its capability in this sector. Because so many sectors of the construction industry can be positively impacted by using VR technology, further research is recommended into the wider adoptability of VR technologies in the industry.

Author details

Mohsen Ghobadi and Samad M.E. Sepasgozar*
Faculty of Built Environment, The University of New South Wales, Australia

*Address all correspondence to: samad.sepasgozar@gmail.com

References

[1] Laurell C, Sandström C, Berthold A, Larsson DJ. Exploring barriers to adoption of Virtual Reality through Social Media Analytics and Machine Learning—An assessment of technology, network, price and trialability. Journal of Business Research. 2019;**100**:469-474

[2] Shen H, Zhang J, Yang B, Jia B. Development of an educational virtual reality training system for marine engineers. Computer Applications in Engineering Education. 2019;**27**(3): 580-602

[3] Sherman WR, Craig AB. Understanding Virtual Reality: Interface, Application, and Design. Morgan Kaufmann; 2018

[4] Schneider R. Mixed reality and its future in design education. In: National Conference on the Beginning Design Student. 2018

[5] Kim D, Ko YJ. The impact of virtual reality (VR) technology on sport spectators' flow experience and satisfaction. Computers in Human Behavior. 2019;**93**:346-356

[6] Hall V, Conboy-Hill S, Taylor D. Using virtual reality to provide health care information to people with intellectual disabilities: Acceptability, usability, and potential utility. Journal of Medical Internet Research. 2011;**13** (4):e91

[7] Chang G, Morreale P, Medicherla P. Applications of augmented reality systems in education. In: Society for Information Technology & Teacher Education International Conference; Association for the Advancement of Computing in Education (AACE). 2010

[8] Nee AY, Ong S-K. Virtual and augmented reality applications in manufacturing. IFAC Proceedings Volumes. 2013;**46**(9):15-26

[9] Vahdatikhaki F et al. Beyond data visualization: A context-realistic construction equipment training simulators. Automation in Construction. 2019;**106**:102853

[10] Eiris R, Gheisari M, Esmaeili B. PARS: Using augmented 360-degree panoramas of reality for construction safety training. International Journal of Environmental Research and Public Health. 2018;**15**(11):2452

[11] Jokkaw N, Suteecharuwat P, Weerawetwat PJ. Measurement of construction workers' feeling by virtual environment (VE) technology for guardrail design in high-rise building construction projects. Engineering Journal. 2017;**21**(5):161-177

[12] Kayhani N et al. Utilization of Virtual Reality Visualizations on Heavy Mobile Crane Planning for Modular Construction. International Association for Automation and Robotics in Construction (IAARC); 2018

[13] de Klerk R, Duarte AM, Medeiros DP, Duarte JP, Jorge J, Lopes DS. Usability studies on building early stage architectural models in virtual reality. Automation in Construction. 2019;**103**: 104-116

[14] Boyd DE, Koles B. An introduction to the Special Issue "Virtual Reality in Marketing": Definition, Theory and Practice. Journal of Business Research. 2019;**100**:441-444

[15] Basu A. A brief chronology of Virtual Reality. arXiv preprint arXiv: 191109605. 2019

[16] Hasnain A, Laffont P-Y, Jalil SBA, Buyukburc K, Guillemet P-Y, Wirajaya S, et al. Piezo-actuated varifocal head-mounted displays for virtual and augmented reality. In: Advances in Display Technologies IX. International Society for Optics and Photonics; 2019

[17] Glegg SM, Holsti L, Stanton S, Hanna S, Velikonja D, Ansley B, et al. Evaluating change in virtual reality adoption for brain injury rehabilitation following knowledge translation. Disability and Rehabilitation. Assistive Technology. 2017;**12**(3):217-226

[18] Paulo MM, Rita P, Oliveira T, Moro S. Understanding mobile augmented reality adoption in a consumer context. Journal of Hospitality and Tourism Technology. 2018;**9**(2):142-157

[19] tom Dieck MC, Jung TH. Value of augmented reality at cultural heritage sites: A stakeholder approach. Journal of Destination Marketing & Management. 2017;**6**(2):110-117

[20] Girvan C. What is a virtual world? Definition and classification. Educational Technology Research and Development. 2018;**66**(5):1087-1100

[21] Flavián C, Ibáñez-Sánchez S, Orús C. The impact of virtual, augmented and mixed reality technologies on the customer experience. Journal of Business Research. 2019;**100**:547-560

[22] Mann S, Furness T, Yuan Y, Iorio J, Wang Z. All reality: Virtual, augmented, mixed (x), mediated (x, y), and multimediated reality. 2018. arXiv preprint arXiv:180408386

[23] Baeza E. Applications of Virtual Reality in Construction. 2018

[24] Tsai H-H, Ho X-Y, Chang C-T, Tsai C-Y, Yu P-T, Chiou K-C. An interactive virtual reality application in education for soil and water conservation. In: 2019 International Symposium on Educational Technology (ISET); IEEE. 2019

[25] Pan Z, Cheok AD, Yang H, Zhu J, Shi J. Virtual reality and mixed reality for virtual learning environments. Computers & Graphics. 2006;**30**(1): 20-28

[26] Grasset R, Boissieux L, Gascuel JD, Schmalstieg D. Interactive mediated reality. In: Proceedings of the Sixth Australasian conference on User interface-Volume 40; Australian Computer Society, Inc. 2005

[27] Li X, Yi W, Chi H-L, Wang X, Chan AP. A critical review of virtual and augmented reality (VR/AR) applications in construction safety. Automation in Construction. 2018;**86**:150-162

[28] Siriborvornratanakul T. A study of virtual reality headsets and physiological extension possibilities. In: International Conference on Computational Science and Its Applications; Springer. 2016

[29] Spaeth AB, Khali R. The place of VR technologies in UK architectural practice. Architectural Engineering and Design Management. 2018;**14**(6): 470-487

[30] Bellini H, Chen W, Sugiyama M, Shin M, Alam S, Takayama D. Virtual & augmented reality: Understanding the race for the next computing platform. In: IFLA Trend Report. Vol. 1. 2016

[31] Gleasure R, Feller J. A rift in the ground: Theorizing the evolution of anchor values in crowdfunding communities through the oculus rift case study. Journal of the Association for Information Systems. 2016;**17**(10):708

[32] Jung TH, tom Dieck M. Augmented reality, virtual reality and 3D printing for the co-creation of value for the visitor experience at cultural heritage places. Journal of Place Management and Development. 2017;**10**(2):140-151

[33] Poushneh A, Vasquez-Parraga AZ. Discernible impact of augmented reality on retail customer's experience, satisfaction and willingness to buy. Journal of Retailing and Consumer Services. 2017;**34**:229-234

[34] Ozacar K, Ortakci Y, Kahraman I, Durgut R, Karas IR. A low-cost and lightweight 3D interactive real estate-purposed indoor virtual reality application. ISPRS Annals of Photogrammetry, Remote Sensing and Spatial Information Sciences. 2017;**4**:307

[35] Stinson C, Bowman DA. Feasibility of training athletes for high-pressure situations using virtual reality. IEEE Transactions on Visualisation and Computer Graphics. 2014;**20**(4):606-615

[36] Nguyen M-T, Nguyen H-K, Vo-Lam K-D, Nguyen X-G, Tran M-T. Applying virtual reality in city planning. In: International Conference on Virtual, Augmented and Mixed Reality; Springer. 2016

[37] Zaker R, Coloma E. Virtual reality-integrated workflow in BIM-enabled projects collaboration and design review: A case study. Visualization in Engineering. 2018;**6**(1):4

[38] Juan Y-K, Chen H-H, Chi H-Y. Developing and evaluating a virtual reality-based navigation system for pre-sale housing sales. Applied Sciences. 2018;**8**(6):952

[39] Wu W, Hartless J, Tesei A, Gunji V, Ayer S, London J. Design assessment in virtual and mixed reality environments: Comparison of novices and experts. Journal of Construction Engineering and Management. 2019;**145**(9):04019049

[40] Zhao D, Lucas J. Virtual reality simulation for construction safety promotion. International Journal of Injury Control and Safety Promotion. 2015;**22**(1):57-67

[41] Shi Y et al. Impact assessment of reinforced learning methods on construction workers' fall risk behavior using virtual reality. Automation in Construction. 2019;**104**:197-214

[42] Muhammad AA, Yitmen I, Alizadehsalehi S, Celik T. Adoption of Virtual Reality (VR) for site layout optimization of construction projects. Teknik Dergi. 2019;**31**(2)

[43] Ciuffini AF, Di Cecca C, Ferrise F, Mapelli C, Barella S. Application of virtual/augmented reality in steelmaking plants layout planning and logistics. Metallurgia Italiana. 2016;**2016**(7):5-10

[44] Zollmann S, Hoppe C, Kluckner S, Poglitsch C, Bischof H, Reitmayr G. Augmented reality for construction site monitoring and documentation. Proceedings of the IEEE. 2014;**102**(2):137-154

[45] Alsafouri S, Ayer SK. Mobile augmented reality to influence design and constructability review sessions. Journal of Architectural Engineering. 2019;**25**(3):04019016

[46] Chalhoub J, Ayer SK. Exploring the performance of an augmented reality application for construction layout tasks. Multimedia Tools and Applications. 2019;**78**(24):35075-35098

[47] Ikiz YD, Atici-Ulusu H, Taskapilioglu O, Gunduz T. Usage of augmented reality glasses in automotive industry: Age-related effects on cognitive load. International Journal of Recent Technology and Engineering. 2019;**8**(3):1-6

[48] Hou L, Chi H-L, Tarng W, Chai J, Panuwatwanich K, Wang X. A framework of innovative learning for skill development in complex operational tasks. Automation in Construction. 2017;**83**:29-40

[49] Davis FD. Perceived usefulness, perceived ease of use, and user acceptance of information technology. MIS Quarterly. 1989;**13**(3):319-340

[50] Venkatesh V, Morris MG, Davis GB, Davis FD. User acceptance of information technology: Toward a

unified view. MIS Quarterly. 2003: 425-478

[51] Ullah F, Sepasgozar Samad M, Siddiqui S. An investigation of real estate technology utilization in technologically advanced marketplace. In: 9th International Civil Engineering Congress (ICEC-2017), "Striving Towards Resilient Built Environment". December 2017

[52] Fernandes KJ, Raja V, White A, Tsinopoulos C. Adoption of virtual reality within construction processes: A factor analysis approach. Technovation. 2006;**26**(1):111-120

[53] Stockinger HJ. The future of augmented reality—An Open Delphi study on technology acceptance. International Journal of Technology Management. 2016;**11**(1):55-96

[54] Whyte J. Innovation and users: virtual reality in the construction sector. Construction Management and Economics. 2003;**21**(6):565-572

[55] Silverman D. Interpreting Qualitative Data. SAGE Publications Limited; 2019

[56] Magolda PM. Doing case study research: A practical guide for beginning researchers. Journal of College Student Development. 2007;**48**(1):123-125

[57] Heinonen M, Ojanen V, Hannola L. Adoption of VR and AR technologies in the enterprise. In: ISPIM Conference Proceedings. The International Society for Professional Innovation Management (ISPIM). 2018